Clay Minerals: Roles in Oil and Gas Generation, Drilling, and Enhanced Recovery

Clay Minerals: Roles in Oil and Gas Generation, Drilling, and Enhanced Recovery

Guest Editors

Guanzheng Zhuang
Qiang Li

MDPI Basel • Beijing • Wuhan • Barcelona • Belgrade • Novi Sad • Cluj • Manchester

Guest Editors

Guanzheng Zhuang
School of Environmental Science and Engineering
Guangdong University of Technology
Guangzhou
China

Qiang Li
Petroleum College
China University of Petroleum-Beijing at Karamay
Karamay
China

Editorial Office
MDPI AG
Grosspeteranlage 5
4052 Basel, Switzerland

This is a reprint of the Special Issue, published open access by the journal *Minerals* (ISSN 2075-163X), freely accessible at: https://www.mdpi.com/journal/minerals/special_issues/70GQYHZ1E5.

For citation purposes, cite each article independently as indicated on the article page online and as indicated below:

Lastname, A.A.; Lastname, B.B. Article Title. *Journal Name* **Year**, *Volume Number*, Page Range.

ISBN 978-3-7258-6798-1 (Hbk)
ISBN 978-3-7258-6799-8 (PDF)
https://doi.org/10.3390/books978-3-7258-6799-8

Contents

About the Editors

Guanzheng Zhuang

Guanzheng Zhuang is an Associate Professor at Guangdong University of Technology, specializing in clay minerals and clay-based materials. He earned his PhD in Material Science and Engineering from China University of Geosciences (Beijing) in 2019, during which he spent a year (2017–2018) as a visiting PhD student at Sorbonne University. Following his doctorate, he served as a Postdoctoral Fellow at the Guangzhou Institute of Geochemistry, Chinese Academy of Sciences. Dr. Zhuang has been actively involved in the international clay science community, having served on the Early Career Clay Scientists Committee of the International Association for the Clay Study (2021–2025). He currently serves on the Editorial Board of *Applied Clay Science* (2023–present) and is a member of the Chinese Society for Mineralogy, Petrology, and Geochemistry (2019–present).

Qiang Li

Qiang Li, PhD, received his doctoral degree in Physical Chemistry of Materials from Sorbonne University, France, in 2024. He is currently an Assistant Professor at the Petroleum College of China University of Petroleum (Beijing), Karamay Campus. His current research interests focus on oilfield chemistry and the application of novel functional materials in oil and gas production.

minerals MDPI

Editorial

Editorial for the Special Issue "Clay Minerals: Roles in Oil and Gas Generation, Drilling, and Enhanced Recovery"

Guanzheng Zhuang [1,*] and Qiang Li [2]

1 School of Environmental Science and Engineering, Guangdong University of Technology, Guangzhou 510006, China
2 Faculty of Petroleum, China University of Petroleum-Beijing at Karamay, Karamay 834000, China; liqiang2024@cupk.edu.cn
* Correspondence: zhuangguanzheng@gdut.edu.cn

Clay minerals are fundamental components of sedimentary basins and play a dual role in the petroleum industry according to Zhuang et al. (2024) [1]. On one hand, they act as critical constituents of source rocks, cap rocks, and reservoirs, influencing hydrocarbon generation, migration, and trapping mechanisms [2,3]. On the other hand, their unique physicochemical properties—such as swelling capacity, cation exchange, and large specific surface area—make them indispensable functional materials in drilling fluids and cementing operations [4,5] (contribution 1). However, these same properties can also present significant engineering challenges, including wellbore instability and formation damage [6].

This Special Issue, "Clay Minerals: Roles in Oil and Gas Generation, Drilling, and Enhanced Recovery," was curated to explore the critical interactions between clay minerals and hydrocarbon systems. The collection of nine papers published herein represents a blend of comprehensive reviews and cutting-edge original research. These contributions span the spectrum from fundamental mineralogical characterization of reservoirs to the development of novel clay-based composites for high-performance drilling and cementing applications.

A significant portion of this Special Issue addresses the perennial challenges of drilling in complex formations, particularly regarding rheology control and shale inhibition. As exploration moves toward deeper and harsher environments, the limitations of traditional bentonite-based fluids become apparent. In a comprehensive review, Zhuang et al. (2024) [5] explore the application of nanofibrous clay minerals, specifically palygorskite and sepiolite, in water-based drilling fluids (WBDFs). They highlight how the unique nanostructure of these minerals offers superior salt resistance and thermal stability compared to conventional smectites, proposing them as a new generation of rheological additives. Building on the theme of high-performance additives, Zhang et al. (2025) (contribution 2) present a novel hydrothermal carbon/bentonite composite (HCBC). Their work demonstrates that these composites not only stabilize rheology but also significantly reduce filtration loss and improve lubrication in high-temperature environments by scavenging free radicals, thereby protecting polymers from degradation.

Two contributions focus specifically on maintaining wellbore stability in reactive shale formations. Wang et al. (2024) (contribution 3) synthesized a small-molecule quaternary ammonium salt (TEE-2) via N-alkylation. Their experimental evaluation proves that this inhibitor effectively suppresses the hydration swelling and dispersion of clay minerals through intercalation and adsorption, offering a promising solution for maintaining wellbore integrity. Complementing this, Ai et al. (2024) (contribution 4) provide a review of mechanical inhibitors. Moving beyond chemical inhibition, they discuss the use of physical plugging agents—including carbon-based and mineral-based nanomaterials—to seal microfractures in shale, thereby preventing pressure transmission and fluid invasion.

Understanding the mechanisms of failure is as crucial as preventing them. Gu et al. (2025) (contribution 5) analyze drilling losses in the Mesozoic reservoirs of the Zhengning oilfield. Through a combination of mineralogical analysis and rock mechanics, they identify that the hydration–dispersion of illite/smectite mixed layers, combined with the propagation of natural fractures, are the primary drivers of wellbore instability in this region. Similarly, Fang et al. (2025) (contribution 6) investigate the impact of hydration on the Gulong shale oil reservoir. Using micro-CT and SEM, they reveal a complex trade-off: while hydration initially increases porosity and permeability through clay dispersion, prolonged exposure leads to crack propagation and a reduction in core strength. Their findings regarding the impact of fracturing fluid concentration offer vital theoretical support for balancing reservoir stimulation with long-term stability.

The presence of clay minerals significantly complicates the petrophysical evaluation of reservoirs. Two articles in this issue address the difficulties in interpreting logging data in clay-rich formations. Yu et al. (2025) (contribution 7) investigate the phenomenon of high resistivity in the ultra-deep shale of the Mahu Sag. By characterizing the mineral composition, they reveal that the reservoir is dominated by brittle minerals (quartz/feldspar) with low clay content. Their findings clarify that high resistivity is driven by the mineral framework and porosity characteristics, providing a theoretical basis for better drillability prediction.

In a detailed review, Navarro-Perez et al. (2025) (contribution 8) examine the petrophysical controls of greensand reservoirs, which are rich in the iron-bearing clay mineral glauconite. They highlight how the paramagnetic and conductive nature of glauconite distorts conventional well logs (such as NMR and resistivity), leading to biased porosity and saturation estimates. The review suggests integrated workflows involving core analysis and spectral gamma-ray logging to overcome these interpretation challenges.

Finally, the utility of clay minerals extends beyond drilling fluids into well construction. Ventura et al. (2025) (contribution 9) evaluate the use of palygorskite as an extender agent in lightweight cement pastes. Their performance analysis indicates that palygorskite-based pastes exhibit lower viscosity and yield stress compared to bentonite-based pastes while maintaining satisfactory stability. This suggests that fibrous clays can be effective alternatives for cementing operations in formations with low fracture gradients.

The research presented in this Special Issue underscores the versatility and complexity of clay minerals in the energy sector. From acting as the root cause of wellbore instability to serving as the solution in the form of advanced fluid additives, clay minerals remain at the heart of petroleum engineering challenges.

We hope this collection provides valuable insights for researchers and engineers working on drilling fluid formulation, reservoir characterization, and well construction. We sincerely thank the authors for their high-quality contributions and the reviewers for their constructive feedback, which ensured the scientific rigor of this Special Issue.

Conflicts of Interest: The authors declare no conflict of interest.

List of Contributions:

1. Zhuang, G.; Zhang, J.; Chen, J.; Liu, Q.; Fan, W.; Li, Q. Application of Nanofibrous Clay Minerals in Water-Based Drilling Fluids: Principles, Methods, and Challenges. *Minerals* **2024**, *14*, 842.
2. Zhang, Y.; Li, D.; Wang, X.; Chen, C.; Zhong, H. Application of Hydrothermal Carbon/Bentonite Composites in Improving the Thermal Stability, Filtration, and Lubrication of Water-Based Drilling Fluids. *Minerals* **2025**, *15*, 1263.

https://doi.org/10.3390/min16010042

3. Wang, Q.; He, H.; Zhao, Y.; Rui, J.; Jiang, R.; Slaný, M.; Chen, G.; Gu, X. Preparation and Performance Evaluation of Small-Molecule Ammonium as a Shale Hydration Inhibitor. *Minerals* **2024**, *14*, 1117.
4. Ai, E.; Li, Q.; Liu, Z.; Zhang, B.; Wang, L.; Ma, C. Advanced Trends in Shale Mechanical Inhibitors for Enhanced Wellbore Stability in Water-Based Drilling Fluids. *Minerals* **2024**, *14*, 1197.
5. Gu, T.; Yang, S.; Xiao, Y.; Zhang, L.; Peng, F.; Su, X.; Fan, T.; Wang, H.; Zhou, D. Analyzing the Mechanism of Drilling Losses in the Zhengning Oilfield in the Ordos Basin. *Minerals* **2025**, *15*, 85.
6. Fang, F.; Xu, K.; Zhang, Y.; Wang, Y.; Xu, Z.; He, S.; Huang, H.; Wang, H.; Jin, W.; Gong, Y. Influence of Hydration on Shale Reservoirs: A Case Study of Gulong Shale Oil. *Minerals* **2025**, *15*, 878.
7. Yu, Y.; Zhang, D.; Zhang, P.; Mu, Z.; Tian, S.; Tan, Y.; Zhou, R. Mineral Characterization and High Resistivity Analysis of Ultra-Deep Shale from Mahu Sag, China. *Minerals* **2025**, *15*, 1171.
8. Navarro-Perez, D.; Fisher, Q.; Lorinczi, P.; Velásquez Arauna, A.; Valderrama Puerto, J. A Review on Greensand Reservoirs' Petrophysical Controls. *Minerals* **2025**, *15*, 1280.
9. Ventura, R.A.; Carvalho, J.V.A.; da Silva, R.R.; Pinto, F.G.H.S.; Freitas, J.C.O.; Pergher, S.B.C.; Palygorskite as an Extender Agent in Light Cement Pastes for Oil Wells: Performance Analysis. *Minerals* **2025**, *15*, 637.

References

1. Zhuang, G.; Li, Q.; Bergaya, F.; Yuan, P. Chapter 1—The significance of clay minerals in drilling and drilling fluids. In *Clay Science in Drilling and Drilling Fluids*; Zhuang, G., Yuan, P., Eds.; Elsevier: Amsterdam, The Netherlands, 2024; pp. 1–19.
2. Li, H.; Hu, Q.; Jones, S.; Gluyas, J.; Ansah, E.O.; Menacherry, S.; Wang, Q.; Ye, T. A review and discussion on the influences of grain-coating clay minerals on water-rock interactions in sandstones. *Earth-Sci. Rev.* **2025**, *263*, 105073. [CrossRef]
3. Wu, L.M.; Zhou, C.H.; Keeling, J.; Tong, D.S.; Yu, W.H. Towards an understanding of the role of clay minerals in crude oil formation, migration and accumulation. *Earth-Sci. Rev.* **2012**, *115*, 373–386. [CrossRef]
4. Zhuang, G.; Zhang, Z.; Jaber, M. Organoclays used as colloidal and rheological additives in oil-based drilling fluids: An overview. *Appl. Clay Sci.* **2019**, *177*, 63–81. [CrossRef]
5. Fan, W.X.; Zhuang, G.Z.; Li, Q.; Yuan, P.; Liu, D. Review of Nanoparticles in Water-Based Drilling Fluids: Innovations, Challenges, and Future Directions. *Energy Fuels* **2025**, *39*, 8800–8826. [CrossRef]
6. Li, Q.; Zhuang, G.; Yuan, P.; Bergaya, F. Chapter 12—Future challenges related to clay minerals in drilling and drilling fluids. In *Clay Science in Drilling and Drilling Fluids*; Zhuang, G., Yuan, P., Eds.; Elsevier: Amsterdam, The Netherlands, 2024; pp. 313–338.

minerals

MDPI

Review

A Review on Greensand Reservoirs' Petrophysical Controls

Daniela Navarro-Perez [1,*], Quentin Fisher [2], Piroska Lorinczi [2], Aníbal Velásquez Arauna [3] and Jose Valderrama Puerto [3]

1 Departamento Ingeniería Química, Universidad de Magallanes, Punta Arenas 6210427, Chile
2 School of Earth and Environment, University of Leeds, Leeds LS2 9JT, UK
3 Gerencia de Exploraciones y Desarrollo, ENAP Magallanes, José Nogueira 1101, Punta Arenas 6200988, Chile
* Correspondence: daniela.navarro@umag.cl

Abstract

This review provides a comprehensive analysis of the petrophysical controls influencing greensand reservoirs, with an emphasis on the role of glauconite and associated clay minerals in determining porosity, permeability, and water saturation. Greensands contain iron-rich clay minerals that exert paramagnetic and conductive effects, challenging conventional well-log interpretations and often leading to biased estimates of reservoir parameters. Several challenges for petrophysical property measurements are faced in the laboratory due to clay-induced pore-throat obstruction and microporosity, which underscores the importance of tailored interpretation workflows and data integration. In this paper we highlight the necessity of integrated approaches such as combining core analysis, spectral gamma-ray, and nuclear magnetic resonance (NMR) logging with conventional well logs to calibrate petrophysical models using shale–sand water saturation models, such as Waxman–Smits and Simandoux, to better characterise economical pay zones. Finally, future research directions are indicated, which include refining the calibration of saturation and permeability models, advancing rock-typing methodologies, and understanding mineralogical influences on reservoir quality to optimise hydrocarbon recovery from greensand reservoirs.

Keywords: greensand reservoir; glauconite; clay minerals; petrophysical properties

1. Introduction

Greensand reservoirs are present in mature basins and marginal plays, where reliable petrophysical evaluation is essential to reduce uncertainty in reserve calculations and optimising production planning. Greensands—or glauconitic sandstones—are a complex rock type that produce oil or gas throughout the world, including the North Sea region of northern Europe, New Zealand, India, Brazil, Argentina, and Chile (see Table 1). These siliciclastic rocks are distinguished due to the predominant presence of glauconite along with other clay minerals such as illite–smectite mixed layers and chlorite in minor proportions. Often greensands have poor to moderate porosity (<15% v/v) and ultra-low to low permeability, classifying them as tight reservoirs, making hydraulic fracturing the only viable technology to make them economically producible.

In addition to their relevance as hydrocarbon-bearing formations, greensands are of increasing interest within the mineralogical community due to their high content of the clay mineral glauconite. The petrophysical behavior of these reservoirs is fundamentally governed by the mineralogical composition, diagenetic evolution, and microstructural characteristics of glauconite and associated clays. These minerals influence key reservoir

 https://doi.org/10.3390/min15121280

properties, including porosity, permeability, and water saturation, through their surface area, cation-exchange capacity, and microporosity. This review aligns with the scope of minerals by highlighting the mineralogical controls on reservoir quality and by emphasising the role of clay minerals in subsurface energy systems, including their implications for hydrocarbon production, CO_2 storage, and even agricultural applications.

The term "greensand" stems from the greenish colour of the rock, which is primarily attributed to the presence of glauconite but is also enhanced by other clay minerals such as chlorite and illite–smectite [1–5]. The presence of these clay minerals in greensands results in several unique properties, such as a bimodal pore-size distribution due to the presence of macropores between framework grains and micropores between clay minerals [1]. Chemical analysis suggests that they often have a high iron, potassium [6], and boron content [7]. They also have a moderate cation-exchange capacity (7.8 to 32.4 MEQ/100 g) [8] due to the high specific surface area owing to the presence of mixed clay minerals. These features significantly affect commonly used wire-line log measurements (e.g., gamma-ray, density, neutron, and resistivity measurements), making standard well-log interpretation techniques unreliable for estimating porosity, clay volume, and water saturation.

Archie's [9] standard water saturation equation remains widely used to estimate hydrocarbon volumes in clay-bearing sandstones, despite being formulated for clean sandstones without conductive clay minerals. In reservoirs like greensands, the assumptions behind Archie's model break down, particularly in conditions of low water salinity or in rocks with high clay content (e.g., [10]). This challenge is part of the "shaly sand problem," where clay minerals distort resistivity-based saturation estimates. Glauconite, in particular, complicates interpretation due to its dual porosity and elevated iron content, which significantly increases rock conductivity. These properties result in low resistivity readings and misleading water saturation estimates. While shale–sand water saturation models such as the Waxman–Smits [11] and the dual-water models [12] attempt to incorporate the effects of clay conductivity, their application is still limited, and their parameters are not well calibrated for glauconitic systems. The selection of a suitable water saturation model for greensand reservoirs is still debated, especially when logs are acquired in low-salinity environments or reservoirs with mixed clay mineralogy.

Net pay is one of the most valuable outcomes of petrophysical interpretation and evaluation; however, greensand reservoirs are inherently complex. Due to their heterogeneous mineralogy, laboratory services confront issues in routine core analysis (RCA) programs, delaying petrophysical and petrological measurements; hence, in many fields, interpreters rely heavily on logging tools. In addition, the cut-off criteria for porosity, clay volume, and water saturation are often inconsistent, and applying fixed thresholds often leads to poor reservoir characterisation. Advanced tools like nuclear magnetic resonance (NMR) and mercury injection capillary pressure (MICP) have improved estimates of fluid saturation and permeability by probing pore structure more directly [13]. These methods also enable cross-calibration (e.g., using NMR T_2 distributions to infer pore-body sizes and MICP to define pore throats), but their integration into consistent workflows tailored to greensands remains underdeveloped.

A few published studies have proposed petrophysical models and workflows for greensand reservoirs (e.g., [8,14–19]); however, these are case-specific and do not fully address the underlying physical mechanisms controlling petrophysical behaviour in greensands. However, a systematic literature review on glauconite research undertaken by Nong et al. [20] indicates that research prospects are increasing, with a special focus on fertilisation, composition features, and provenance as emerging frontiers. This trend is further supported by recent advances in glauconite applications in agriculture, as reviewed by Dasi et al. [21], highlighting the mineral's broader relevance beyond subsurface energy systems.

This review paper critically examines the petrophysical properties of greensand reservoirs, focusing on

- Key physical and mineralogical controls on porosity, permeability, and irreducible water saturation;
- Measurement and interpretation challenges, particularly in log analysis and saturation modelling;
- Recent advances in core–log integration and workflow development for more reliable reservoir characterisation.

By synthesising the available literature in English and Spanish and identifying the limitations of current models and assumptions, this review contributes to the development of more appropriate interpretation frameworks. It emphasises that many conventional petrophysical tools and models—particularly those based on clean sandstones—are inadequate for glauconitic systems. The review concludes with recommendations for integrated interpretation workflows that combine advanced logging methods, core and laboratory calibration, and tailored petrophysical models—incorporating mineralogical complexity, adjusted NMR parameters, and electrochemical analysis—while emphasising the need for future research in core–log–digital rock physics integration and machine learning approaches to improve reservoir characterisation and management in greensand reservoirs.

Table 1. Porosity and permeability data reported from some greensand reservoirs worldwide.

Reservoir Type	Reservoir Name	Location	Geological Period or Epoch	Porosity * (%v/v)	Permeability * (mD)	Glauconite Content * (%v/v)	Ref.
Oil	Matulla Formation from Nezzazat Group	Lower Senonian, Gulf of Suez, Egypt	Upper Cretaceous	3.5 to 25	<1 to 90	20 to 45	[8]
Oil and gas	Glauconite member from Mannville Group	Lake Newell, Southern Alberta, Canada	Lower Cretaceous	9 to 33	10 to 10,200	Not available	[22]
Tight oil	Albian "A" Sand Formation from Greensand Group	North Celtic Sea Basin	Lower Cretaceous	≤30	≤50	≤45	[23]
Oil	Pona Reservoir, Chonta Formation	Peruvian forest region	Cretaceous	1 to 20	1 to 1000	Not available	[24]
Tight oil	Caballos Formation	Putumayo Basin, Colombia	Lower Cretaceous	2 to 19	0.01 to 1200	10 to 60	[25]
Tight oil	Mardie Greensand Formation	Carnarvon Basin, Australia	Early Cretaceous	15 to 28	0 to 100	Not available	[14,15]
Oil	Hermod and Ty Formations	Siri Canyon, Danish North Sea	Palaeocene	25 to 40	60 to 1000	20 to 30	[26]
Tight gas	Magallanes Formation	Campo Indio field, Austral Basin, Argentina	Upper Cretaceous	15 to 35	0.01 to 1	Not able to be determined **	[9,10]
Tight gas	Zona Glauconitica	Austral-Magallanes Basin, Chile	Palaeocene	10 to 25	0.001 to 1	Not able to be determined **	[18]
Tight oil	Upper T of Napo Formation	Oriente Basin, Ecuador	Cretaceous	5 to 20	≤500	10 to 40	[27]

* Reported either from core analysis or reservoir evaluation. ** XRD analysis could not identify glauconite since it overlapped with illite presence.

2. Geological and Mineral Framework of Greensands

2.1. Mineral Composition and Diagenesis

This subsection outlines the primary mineral constituents and the process of glauconitisation that governs greensand reservoirs, with an emphasis on how mineral evolution influences petrophysical behaviour.

Greensands are iron-rich sandstones dominated by the clay mineral glauconite, a member of the mica group. Other clay minerals such as chlorite, smectite, and illite are frequently present [1–5]. Glauconitic pellets form through a process called glauconitisation, in which K-poor smectite transforms into K-rich glauconite mica pellets during burial diagenesis [6].

As glauconitisation advances, potassium and iron contents increase within these pellets, while aluminium content decreases (Figure 1). The resulting sedimentary rock comprises a complex mixture of clay minerals, with diagenetic alterations that directly affect the texture and reservoir quality of greensands [6]. For example, transformations from smectite to illite or glauconite alter the cation-exchange capacity and permeability. The development of clay mineral coatings or pore-filling cements during diagenesis modifies both the intergranular cohesion and the pore-throat geometry, significantly affecting porosity and increasing surface area. Other clay minerals—chlorite, smectite, and illite—also play significant roles in the mineralogical and petrophysical behaviour of greensands. Chlorite generally occurs as a pore-filling and grain-coating material, whereas smectite and illite commonly exist as fine matrix minerals or mixed-layer clays (see, e.g., [1,2]). As shown in Figure 2, these minerals often act as grain coating or pore-filling phases, contributing to microporosity and reduced permeability.

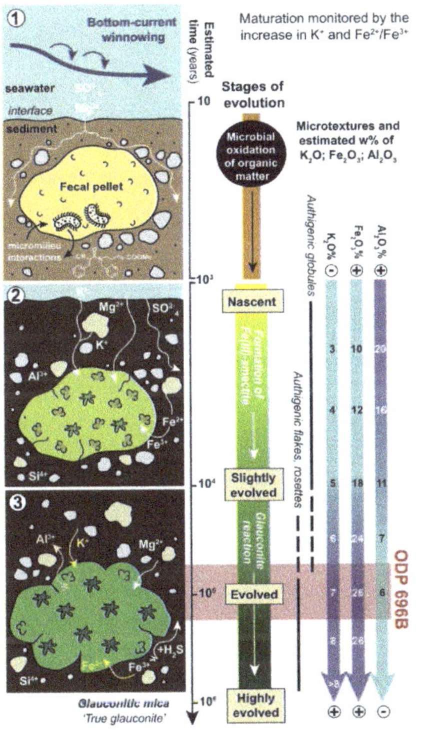

Figure 1. Glauconitisation model in the studied grains (modified after López-Quirós et al. [28], following the concepts of Odin and Matter [29] and Baldermann et al. [30]). Adapted from López-Quirós et al. [31]. Nascent stage: After deposition at the seafloor, glauconite grains begin forming within semi-confined environments. Early pore networks start developing alongside the incorporation of detrital clay minerals. During this stage, dissolution of carbonate components occurs, and iron is primarily present as Fe^{3+}-smectite precursors. Microbial oxidation of organic matter contributes to the early geochemical conditions, typically over a period of ~10–10^3 years. Slightly evolved stage: The grains mature as ions from the formation water re-accommodated within the mineral structure, promoting the neoformation of glauconitic smectite (G/S). Detrital minerals gradually diminish,

coinciding with an increase in Fe and K ion content and the transformation of iron oxides to Fe^{2+} and Fe^{3+} forms. This stage generally spans ~10^3–10^4 years. Evolved stage: Continued recrystallisation and rearrangement of mineral composition result in more consolidated grains with the development of superficial cracks. Crystal growth accelerates particularly in the centre of the grain. The chemical composition evolves significantly as Si, Al in octahedral coordination (Al^{VI}), Mg, Ca, and Na are progressively substituted by Al in tetrahedral coordination (Al^{IV}), Fe (primarily Fe^{2+}), and K. This reaction from Fe(III)–smectite to glauconite occurs over a period of between ~10^4 and 10^6 years. Highly evolved stage: At this final stage, a new generation of glauconitic minerals forms on the grain periphery, encapsulating the original glaucony pellets. This process involves advanced recrystallisation, leading to the development of glauconitic micas with distinctive microtextures, such as authigenic flakes and rosettes, indicative of full maturation.

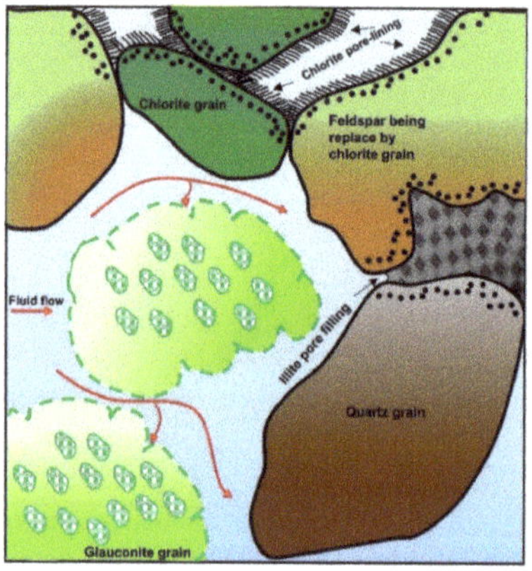

Figure 2. Diagram of greensand major grains and clay distribution types as grain-coating and -filling pore throats; glauconite exhibits microporosity [32].

2.2. Diagenetic Alteration and Textural Controls

Beyond mineral composition, the diagenetic evolution of greensands plays a critical role in shaping reservoir quality. This subsection examines grain morphology, cementation, and depositional textures that influence pore structure and connectivity.

Glauconite grains exhibit diverse morphologies, including spherical, ovoid, tabular, vermiform, and lobate shapes, and may be incorporated within microfossils [33]. These grains typically tend to be part of the rock framework rather than acting as a pore-filling phase, thereby reducing the pore-throat area and connectivity. Silica cement is typically present in glauconitic sandstones that strengthen the pore network but reduce the overall porosity. Carbonate (e.g., calcite) or berthierine cement may also occur locally, blocking intergranular pore throats. Some authors describe the glauconitic grains as having angular to subangular (i.e., irregular) shapes with highly rough surfaces. Sorting varies from locally poor to moderate, while packing ranges from moderately poor to a tight arrangement (e.g., [27,34–36]). Therefore, reservoir quality in greensands is principally controlled by their characteristic poor sorting, rough grain surfaces, and tight grain packing—features inherited from depositional conditions and modified by diagenesis. These textural heterogeneities lead to complex pore networks with restricted pore-throat sizes and reduced connectivity, especially where clay minerals are abundant.

These textural features are not only the result of diagenesis processes but are also inherited from the original depositional environment. As shown in Figure 1, greensand formation is closely tied to specific depositional environments, typically low-energy settings with slow sedimentation rates, such as outer shelf or upper slope conditions. These environments favour the authigenic development of glauconite through prolonged residence times at the sediment–water interface (i.e., glauconitisation), allowing gradual transformation of smectite precursors into mature glauconitic pellets [28,29]. The depositional setting also influences grain sorting, packing, and the abundance of detrital versus authigenic clay minerals, which in turn affect reservoir quality. Poor sorting and tight grain arrangements inherited from these environments contribute to reduced pore-throat connectivity. This, in turn, increases microporosity, as further described in the next subsection. Consequently, the depositional environment exerts a first-order control on the petrophysical behaviour of greensands, particularly porosity, permeability, and irreducible water saturation (Sections 3.2–3.4), which are essential for classifying greensand reservoir quality and guiding petrophysical model calibration. Finally, because glauconite functions simultaneously as a framework grain and a conductive clay phase, many standard petrophysical models that assume a strict partitioning between sand and clay fractions are not fully applicable to greensands.

2.3. Microstructure and Dual Porosity

The pore architecture of greensands is notably complex, characterised by a dual porosity system [1,34]. This subsection explores the distribution of macropores and micropores, their petrophysical implications, and challenges for reservoir evaluation.

Macroporosity corresponds to intergranular pore spaces between framework grains, whilst microporosity is mainly controlled by the presence of clay minerals—especially glauconite and chlorite—that create fine networks of isolated pores [34]. Microporosity is generally defined as pores smaller than 0.5 μm [37] and is linked to the high surface area contributed by detrital and authigenic clay minerals.

Figure 3 presents backscattered electron (BSE) images of two greensand samples, with a notable scale bar of 1 μm for micropores. The coexistence of macropores and micropores results in a bimodal pore-throat size distribution (PSD) (Figure 4). These micropores retain bound water, causing high irreducible water saturation [38,39]. This phenomenon presents critical implications for reservoir performance, including fluid storage capacity, flow characteristics, and hydrocarbon recovery strategies.

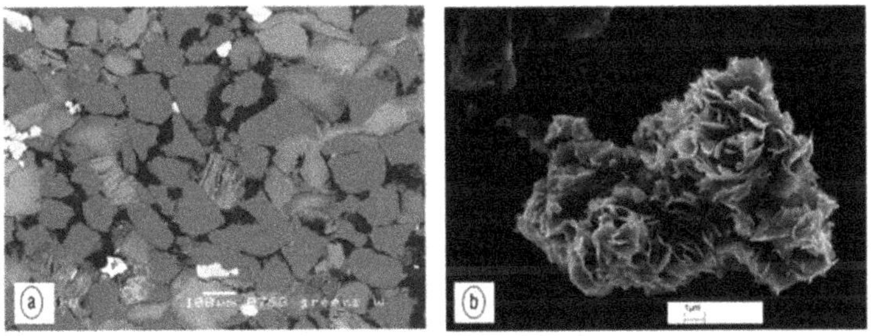

Figure 3. (**a**) BSE image of the North Sea greensand; scale bar for greensand is 100 μm, and the image represents macroporosity. Q, quartz; Gl, glauconite; M, mica. (**b**) BSE glauconite grain image from Arnager greensand; scale bar for glauconite grain is 1 μm. Micropores reside within glauconite grain. Source: Hossain et al. [34]. © Society of Exploration Geophysicists (SEG). Reproduced with permission. DOI: https://doi.org/10.1190/1.3064151.

Because micropores are often isolated from the macropore network, effective porosity is reduced due to poor pore interconnectivity. However, this pore structure tends to preferentially retain water over hydrocarbons (a water-wet rock), which renders greensands economically viable reservoirs. From a pore-volume standpoint, microporosity largely controls key petrophysical properties such as capillary pressure, porosity, permeability, and water saturation. Additionally, abundant dispersed, laminated, or structurally incorporated clays commonly obstruct pore throats, further reducing effective porosity. This contributes to highly scattered and atypical porosity–permeability trends that differ significantly from those observed in clean sandstones (Figure 5). Such complexity poses considerable challenges for the evaluation and modelling of greensand reservoirs, requiring tailored workflows and calibration strategies to accurately assess reservoir quality.

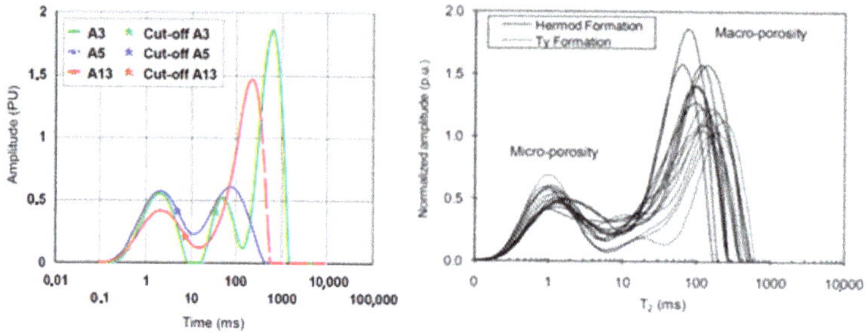

(a) (b)

Figure 4. NMR T_2 curve distribution for greensand samples: (**a**) T_2 spectra for three Well-A core plugs from a Late Paleocene/Early Eocene sandstone, North Sea [2]; (**b**) Nini field, North Sea [40]. © Society of Exploration Geophysicists (SEG). Reproduced with permission. DOI: https://doi.org/10.1190/geo2010-0366.1. A peak close to 1 ms represents microporosity, and a peak close to 100 ms represents macroporosity.

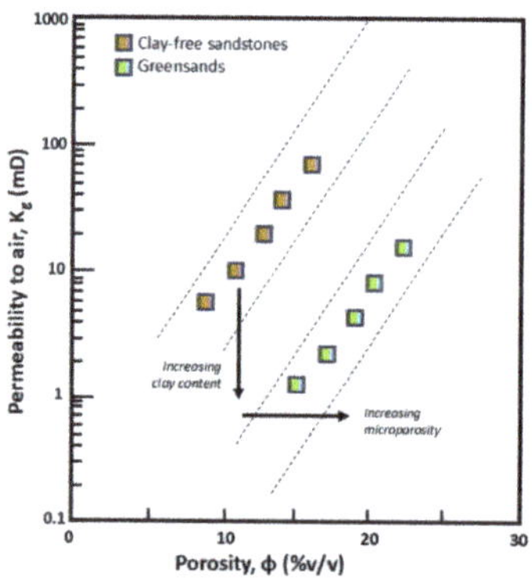

Figure 5. Total porosity and air permeability cross-plot diagram showing the different trends of clay-free sandstones and greensands [32].

3. Greensand Measurement Challenges

3.1. Mineralogical Analysis

Petrographic analysis using transmitted and reflected light microscopy remains a fundamental tool in the mineralogical characterisation of reservoir rocks. This technique enables geologists to visually examine mineral grains, textures, and their spatial relationships at the microscopic scale. It provides direct insights into the identification of minerals, grain morphology, and diagenetic features, thereby complementing bulk analytical techniques such as X-ray diffraction (XRD). In particular, petrographic observations allow the discrimination of clay minerals based on optical properties such as birefringence, pleochroism, and interference colours, which are often indistinguishable by XRD alone. Furthermore, ternary diagrams plotting the relative proportions of key clay minerals—such as glauconite, chlorite, and smectite—are widely used to classify mineral assemblages and interpret diagenetic pathways in greensand reservoirs [31]. Such a mineralogical framework, integrating petrographic observations and quantitative analyses, supports the evaluation of reservoir quality and geological history by quantifying the relative abundance and interplay between these clay minerals. However, due to the greenish colour imparted by glauconite, these sandstones may be misidentified as containing other clays, such as odinite, clinochlore, or chamosite (all from the chlorite group), and berthierine (from the kaolin group). These minerals like glauconite, typically develop at a slow sedimentary deposition rate [41].

In addition, glauconite can be difficult to identify precisely by XRD because its basal reflections often overlap with those of other 10 Å lattice clay minerals, such as illite and mixed-layer clays [42–47]. Recent clay XRD studies have demonstrated that the deconvolution of ethylene glycol-treated samples and the detailed analysis of basal reflections provide a reliable means to distinguish glauconite from mixed-layer silicates, revealing its R3-ordered glauconite–smectite structure with minor expandable layers [31]. Therefore, XRD software tools must be appropriately calibrated to apply such deconvolution techniques effectively and prevent misinterpretation of clay minerals.

3.2. Porosity

Common porosity measurement methods include gas porosimetry based on Boyle's law—typically using helium or nitrogen; brine porosity, determined by the Archimedes' method using fully brine-saturated core samples [48]; and nuclear magnetic resonance (NMR) techniques, which measure the transverse relaxation time (T_2) decay [49]. Gas porosimetry accurately measures both connected and microporous pore volumes due to the small molecular size of the gas, although some isolated pores may remain undetected. In contrast, brine porosity measurements can be influenced by the presence of clay-bound and irreducible water [48]. NMR porosity measurements are increasingly used because they can discriminate between fluid types within pore spaces and are non-invasive [50]. Moreover, laboratory technicians should report a drying protocol and a test for residual bound water and discuss the potential underestimation of pore volume when conventional drying is assumed to be complete.

Porosity measurement in greensand reservoirs is complicated by the presence of iron-rich clay minerals and heterogeneous pore systems. Hossain et al. [40] conclude that helium core porosity effectively measures total porosity because the molecular size of this inert gas is small enough to access the micropore network and clay-bound water. In contrast, the Archimedes method, applied to fully brine-saturated core plugs, measures effective porosity since the water molecule is larger than helium and cannot penetrate tightly bound pores. Their core NMR porosity closely matched the brine porosity, with an average absolute error of 2.88 porosity units (p.u.), and was significantly lower than the core helium porosity, with an average absolute error of 3.77 p.u. This finding matches Rueslatten et al.'s [2] studies on oil greensands, since glauconite is paramagnetic due to its iron content, which generates a shortened T_2 relaxation time curve distribution, and it can also lead to surface relaxivity miscalculations for converting the NMR data to a PSD.

Because iron-bearing glauconite produces local magnetic-susceptibility contrasts and strong internal field gradients, empirical adjustments to T_2 cut-off values have to be considered—typically between 10 and 30 milliseconds for oil reservoirs, according to Dodge et al. [1]—which is significantly lower than the conventional sandstone T_2 cut-off of 33 milliseconds [50], but applying these lower thresholds without independent validation risks misclassification of bound versus free fluids when susceptibility-driven relaxation dominates. Empirical T_2 thresholds can mask fundamental signal distortion caused by inhomogeneous magnetic susceptibility; therefore, reliance on such corrections should be stated explicitly as provisional pending physics-based models.

To reduce ambiguity and increase confidence in NMR-based porosity and PSD interpretations of glauconitic greensands, we recommend the following workflow: measure and report bulk magnetic susceptibility for core plugs and cuttings; report full NMR acquisition parameters and justify any T_2 threshold or relaxivity adjustments; calibrate NMR-derived porosity and PSD against independent pore-scale measurements (helium porosity, brine-saturated/Archimedes porosity, MIP, μCT, and core NMR) and report calibration metrics; use complementary tests (comparison of brine and helium porosities, MIP/μCT constraints, and electrical conductivity measurements) to distinguish bound from free fluids; and, where feasible, adopt or test physics-based, susceptibility-aware NMR approaches (field-dependent relaxometry, susceptibility-corrected models, or joint inversions) rather than relying solely on empirical T_2 cut-offs—these steps document the limits of empirical corrections and provide a pathway to more physically consistent NMR interpretations in greensand systems.

3.3. Permeability

Permeability in greensands exhibits highly scattered trends when plotted against porosity, reflecting the complex pore structure and clay distribution (Figure 5). The presence of mixed clay minerals—especially glauconite and chlorite—substantially affects permeability by coating grain surfaces and plugging pore throats, thus reducing interconnected flow pathways. The pore-throat space of a rock can be measured with the specific surface area (SSA), and clay minerals generally display moderate to high SSA values (see Table 2), which correlate with their greater microporosity and lower permeability. Hossain et al. [40] reported a range of SSA from 17 to 23 m^2/g through nitrogen adsorption (the SSA of quartz is ~2.5 m^2/g) and found an inverse relationship with the macroporosity and permeability. Hossain and Zhou [51] show a relationship between porosity and permeability, with the SSA discriminated by the greensand's pore-filling clay distribution.

Table 2. Cation-exchange capacity range and total surface area for clay minerals (from Navarro-Perez et al. [52]; adapted from Van Olphen and Fripiat [53]; McPhee et al. [48]).

Clay Mineral Group	CEC (MEQ/100 g)	Surface Area (m^2/g) [1]
Smectite	70–130	800
Illite	25–40	30
Chlorite	10–40	15
Kaolinite	3–15	15

[1] Summation of internal (interlayer) and external surface area.

Several researchers have developed innovative methodologies to determine the permeability of greensand in the subsurface by calibrating well-log data using core measurements. The discrimination of iron content, rock matrix density, and NMR parameters is important to define an optimised permeability prediction. Table 3 summarises the main permeability equations in the literature used for greensands. The Kozeny, Timur–Coates, and SDR equations are most frequently applied, either in their classical or modified style. The latter two, NMR-based permeability models, have been developed to integrate relaxation and diffusion properties; these approaches show promise for more reliable permeability estimation in clay-rich reservoirs by accounting for pore-size heterogeneity and clay influence. Further information on the methods for permeability estimation for greensands in the literature is displayed in Table 4. Moreover, permeability predictions based on total SSA must be reconsidered in glauconitic systems because a portion of the surface is associated with hydraulically ineffective microporosity and bound water; models using Kozeny require filtering or redefinitions in their inputs.

Table 3. Permeability equations applied for greensand reservoirs.

Name	Equation	Reference
Timur and Coates (TC)	$k = (\phi_{NMR})^4 (FFI/BVI)^2$	[54]
Schlumberger–Doll (SRD)	$k = 4.5(\phi_{NMR})^4 (T_{2.g})^2$	[55]
Kozeny	$k = c\frac{\phi^3}{S^2}$	[56]
Modified Kozeny's to include macroporosity range	$k = c\phi\rho_2{}^2 \sum_{i=1}^{N} f_i (T_{2i})^2$	[40]
Worthington's modified Kozeny's equation	$k = \left(\frac{B}{F}\right)^{1/C}$	[57]
Modified Kozeny's as a function of rock type	$k = c\frac{\phi^{3-2a}}{10^{2b}}$	[51]

Nomenclature: k = permeability (mD), ϕ_{NMR} = NMR porosity (v/v), FFI = NMR log free fluid index (v/v), BVI = NMR bulk volume irreducible fluid (v/v), $T_{2.g}$ = geometric mean of T_2 distribution time (ms), c = Kozeny's constant, S = specific surface area of bulk (1/μm), ρ_2 = surface relaxivity (μm/ms), f_i = fraction of the total amplitude of each T_{2i}, F = electrical formation factor ($F = a/\phi^m$); B, C, a, and b are fitting constants.

Table 4. Permeability estimation methods found in the literature review used on greensand reservoirs.

Method	Comments	Ref.
Log NMR permeability with SRD and Timur–Coates (TC) equations.	• Iron content underestimates NMR-derived permeability for the TC equation; care must be taken to define the T_2 cut-off value. • The fitted factors of the formulas used were enough for good calibration. • Suggested T_2 cut-off values according to the iron content in greensand reservoirs.	[1]
Electrofacies pattern linked to permeability transforms algorithm using core porosity and permeability.	• Log synthetic permeabilities used as input were litho-density, neutron, sonic, and gamma-ray spectrometry logs. • Grouping the electrofacies (lithological units) is extremely important.	[58]
NMR permeability with the Timur–Coates equation.	• T2 cut-off values below 10 ms for chloritic greensands and 33 ms when chlorite and glauconite are present (oil production).	[2]
Log NMR permeability with the Timur–Coates equation.	• Chlorite presence influences the irreducible bulk volume when using the TC equation. • Density-derived porosity log replaced the FFI/BVI term in the TC formula, increasing permeability estimation.	[38]
Artificial neural network technique to estimate permeability.	• 147 log response sets were trained with core permeability from eight wells. • The synthetic permeability log shown as an example gave good consistency results.	[15]
Density Constrained Stochastic Modelling (DCSM).	• Method based on a random selection of core bulk density values matched with an apparent matrix density log. • The average density tolerance between the two parameters is less or equal to 0.05 g/cm^3.	[17]
Modified Kozeny's equation to include the whole macroporosity range of T_2 times.	• Kozeny's factor, C = 8.3, was estimated from the Mortensen et al. porosity model. • Surface relaxivity derived from specific surface area from Kozeny's BET and image analyses. • The BET specific surface area takes into account the microporosity range, while the image and Kozeny's specific surface area are associated with the macroporosity range.	[40]
Empirical correlation of permeability with the electrical formation factor—Worthington's modified Kozeny's equation.	• The formation factor was estimated using the equation $F = a/\phi^m$, with $a = 1.67$ and $m = 1.9$. • A linear relationship between permeability and formation factor was found by varying the specific surface area. • The diagenesis of greensands needs to be known to replicate this method.	[57]
Modified Kozeny's equation by including two factors, *a* and *b*, dependent on the rock type, instead of the specific surface area or surface relaxivity, and as a function of porosity.	• Factors a and b are constants derived from a linear relationship between the specific surface area and porosity. • For clean sandstone, a = −1.0 and b = 2.2; for clay-bearing sandstone, a = −2.5 and b = −2.3. • For greensand with pore-filling cementation, a = −5.5 and b = 3.0. • A linear relationship between Kozeny's constant, c, and porosity is proposed.	[51]

3.4. Irreducible Water Saturation

Since greensands have a complex microstructure due to the presence of clay, it is expected that the irreducible water saturation (S_{wirr}) and the capillary pressure curves will be high (Figure 6). The glauconitic grains are water-wet [16], favouring storing water in the micropore range and leaving hydrocarbon in the macropore range (Figure 7). Table 5 summarises reported values for the irreducible water saturation in greensands, with values of up to 78% v/v, according to the literature.

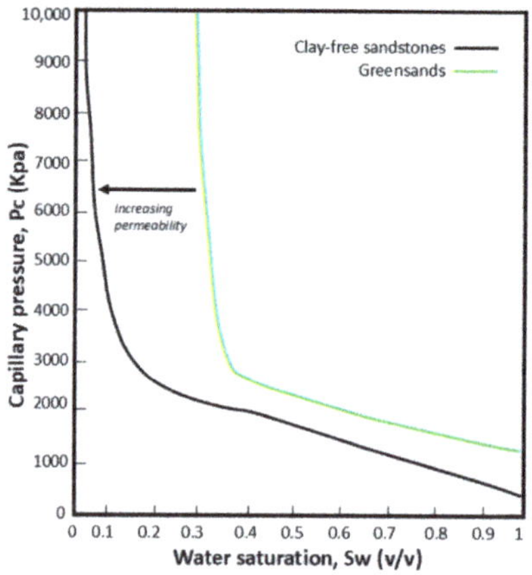

Figure 6. Capillary pressure and water saturation cross-plot comparing clay-free sandstone and greensand trends [32].

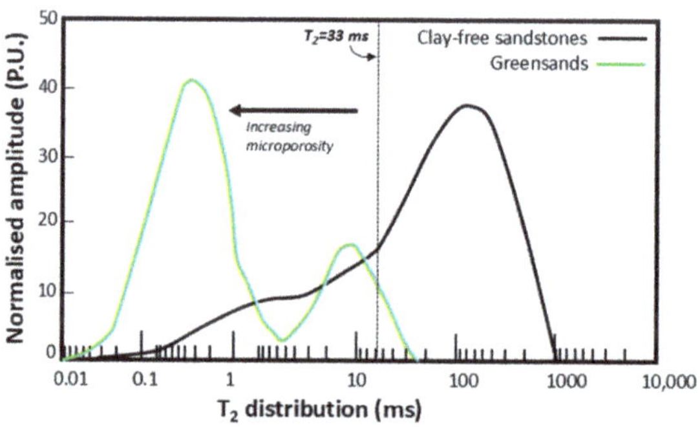

Figure 7. NMR T_2 distribution diagram comparing clay-free sandstone and greensand distributions [32].

Table 5. Irreducible water saturation values found in greensands worldwide along with the technique used to define them.

S_{wirr} Range [1] (%v/v)	T_2 Cut-Off Value (ms)	Technique Used	Ref.
12–78	10 to 30	Centrifuge air/brine drainage capillary pressure and NMR T_2 distribution.	[1]
27–55	33	Decane–brine saturation, Karl Fischer titration, and NMR T_2 distribution	[2]
27–41	33 and 10 for chlorite zones	Karl Fischer titration and NMR T_2 distribution	[38]
22–41	5.2 and 3.7	Centrifuge air/brine drainage capillary pressure and NMR T_2 distribution.	[40]

[1] S_{wirr} is irreducible water saturation.

As previously stated in Section 3.2, the iron content in greensands affects the NMR T_2 readings and hence the estimated S_{wirr}; that is why the authors mentioned in Table 5 used lower T_2 cut-off values. Dodge et al. [1] suggest that the T_2 cut-off value needs to be lower than 10 ms to quantify S_{wirr} [38]. Typical T_2 cut-off values of 33 ms for free mobile fluid and 3 ms for clay-bound water for greensand reservoirs are incorrect for greensands. The T_2 decays faster due to the iron paramagnetic properties and narrower pore-size distribution; thus, the T_2 cut-off value must be lowered (e.g., [59]). Hossain et al. [40] went further, and using the capillary pressure curves, they defined a cut-off value of 100 psig to separate the microporosity from the meso- and macroporosity zone equivalent to a T_2 cut-off value of 5.21 ms.

Furthermore, the gas–water permeability curve for greensands is expected to shift to the right (Figure 8) owing to the high S_{wirr} and its microporosity portion. The water relative permeability curve will start at a moderate to high S_{wirr} and effectively flow from 2% of k_{rw} onwards [60]. In comparison, the gas relative permeability curve will start at 100% k_{rg} and reach a small value of residual gas saturation (1-S_{rg}), where the remaining hydrocarbon gets trapped.

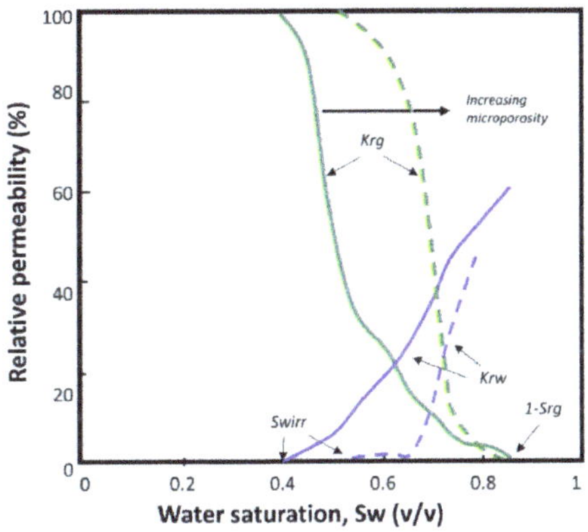

Figure 8. Gas–water relative permeability and water saturation cross-plot comparing the trend of clay-free sandstones (continuous line) and greensands (dashed line). K_{rg} is gas relative permeability, K_{rw} is water relative permeability, S_{wirr} is irreducible water saturation, and S_{rg} is residual gas saturation [32].

3.5. Water Saturation Modelling

Greensands often have high rock conductivity due to the significant cation-exchange capacity of glauconitic grains. Weaver [61] reports values from 5 to 12 meq/100 gr for glauconites with up to 5% expandable clays, and with 50% expandable layers, CEC values reach up to 40 meq/100 gr. Patchett et al. [8] reports values from 7.8 to 32.4 meq/100 gr (13 to 81% glauconite). In contrast, values for greensand deep resistivity logs range from 1 to 5 ohm·m [19,38,62]. Since these values are low, a reservoir can be confused with freshwater rather than a hydrocarbon zone.

Greensands are referred to as low-resistivity pays, according to the definition by Worthington [63] (p.78): "a lack of useful positive contrast in measured electrical resistivity between zones that contain and produce hydrocarbons in commercial quantities and zones that contain and produce only water, within the same reservoir system". Sneider [64] uses the term "low-resistivity, low-contrast" (LRLC) to refer to sandstones with very low resistivities (<2 ohms·m) that have low contrast in resistivity with adjacent shales. Excess clay minerals, fine grains, and microporosity are some of the leading causes of low-resistivity pays, which are features of greensands.

Archie [9], Waxman–Smits [11], and Simandoux [65] are the three models used frequently for greensands worldwide (Tables 6 and 7). Waxman–Smits is the main representative of the double-layer models and uses the cation-exchange capacity parameter. Simandoux represents Vsh models since it adds a shaley term. The Vsh models are based on the wetted-shale/clay volume fraction, meaning that they implicitly assume that the electrical properties of clays, independent of how they are distributed in sandstones, are the same as the layers of the detrital clay adjacent to the sandstones. The advantage of using them is that shale/clay parameters can be estimated from log data alone [10].

Table 6. Water saturation models used on greensands from literature.

Location	Model	Ref.
Trimble Field, MS, USA	Pseudo-Archie using m = 1.8 and n = 1.77	[63]
Nini Field, North Sea	Pseudo-Archie using a = 1.67, m = 1.18, n = 2.4 and R_w = 0.077 ohm·m.	[66]
Trembul Field, Indonesia	Waxman–Smits using a = 1.0, m = 1.76, n = 1.79 and R_w = 0.19 ohm·m @148 °F.	[19]
Glauconite Formation, Southern Chile	Simandoux using a = 1, m = 2.09, n = 1.66.	[67]
Magallanes Formation, Southern Argentina	Pseudo-Archie using a = 0.8–0.82, m = 1.4–1.6, n = 1.4–1.6 and R_w = 0.3 ohm·m @150 °F.	[68]

Table 7. Water saturation models expressed in conductivity terms used in greensands.

Model	Equation	Ref.
Archie	$C_t = \frac{C_w \phi^m S_w{}^n}{a}$	[9,69]
Simandoux	$C_t = \frac{C_w}{F} S_w{}^n + V_{sh} C_{sh}$	[62,65]
Waxman–Smits	$C_t = \frac{C_w}{F^*} S_w{}^n + \frac{BQ_v}{F^*} S_w{}^{n-1}$	[10,11]

Nomenclature: S_w = water saturation (v/v), C_w = water conductivity (S/m), C_t = true rock conductivity (S/m), F = Archie formation factor, a = tortuosity factor constant, ϕ = porosity (v/v), m = Archie cementation exponent, n = Archie saturation exponent, F^* = apparent formation factor, Q_v = cation-exchange capacity per unit pore volume (meq/mL), B = equivalent conductance of clay (mho·ml/(meq·m)), C_{sh} = shale conductivity (S/m), V_{sh} = shale or clay volume (v/v).

The double-layer models offer a better scientific understanding of the rock's electrical behaviour. However, they require core data to calibrate their terms, such as Qv, for log-derived parameters (e.g., [48]). It is important to note that these water saturation models—Archie, Simandoux, and Waxman–Smits—do not use glauconitic sandstones as empirical data for their postulates, so care should be taken when applying them to other rock-type reservoirs such as greensands.

4. Impact of Greensand Well Logs

The interpretation of well logs in greensand reservoirs poses significant challenges due to their complex mineralogical and petrophysical characteristics (see Table 8). A prominent issue is the notably high gamma-ray response, primarily attributed to the elevated potassium content in glauconite. This can be misinterpreted as shaliness, resulting in overestimation of clay volume if standard clay or shale cut-offs are applied. Spectral gamma-ray logs are especially valuable where available. In particular, K/Th ratios combined with the photoelectric factor log can be employed to identify glauconite [70].

Table 8. Summary table showing how greensand affects well-log readings and which petrophysical properties are consequently under- or overestimated.

Well Log	Greensand Effect	Causes	Petrophysical Estimation	Reference
Gamma-ray	Higher response than clean and shaly sandstone.	High presence of potassium. Greensands can contain other clay minerals, such as chlorite, siderite, and pyrite. Higher boron content than other clay minerals.	The shale or clay volume calculated is overestimated.	[17,27,35]
Deep resistivity	Lower response similar to freshwater zones.	High cation cation-exchange capacity (CEC), meaning lower resistivity. Large amount of bound water in the glauconitic grains.	The hydrocarbon zone is not easily identified. The water saturation is overestimated.	[8,35,71]
Neutron porosity	Higher response than sandstone.	Large thermal neutron absorption cross-section because of its high iron content. Higher boron content than other clay minerals.	Neutron porosity is overestimated.	[8,17]
Bulk density	Higher response than sandstone.	Glauconite is denser than sandstone. Iron presence.	Density porosity is underestimated.	[8,17,25,71]
Photoelectric factor	Higher response than sandstone.	Relatively high molecular weight. Iron presence.	It can be used to identify the glauconitic sandstone lithology.	[8,35]
NMR tool	T_2 decays faster.	Paramagnetic effect of the iron content of greensand.	Bulk mobile fluid underestimated. T_2 cut-off value needs to be lower than 30 ms.	[1,2,38]

Due to the greensand effect observed in well logs, core analysis is essential for reliable calibration and petrophysical model development. For example, as shown in Figure 9, the Hermod and Ty formations display the lowest resistivity readings and minimal density–neutron separation, even though core helium porosity and permeability indicate that these intervals remain permeable (permeability greater than 200 mD) and porous (porosity between 38 and 43% v/v).

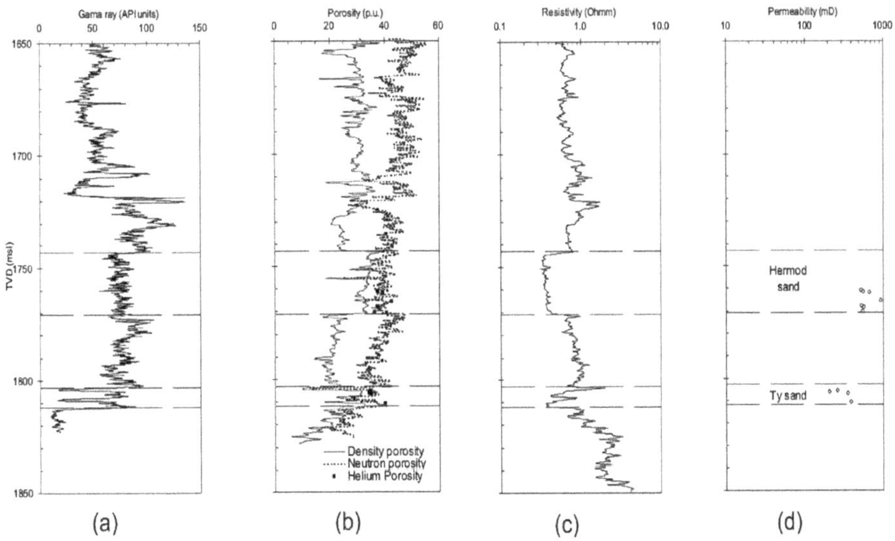

Figure 9. Nini-1 well logs showing the greensand reservoirs intervals (Hermod and Ty sands from the North Sea): (**a**) gamma-ray log; (**b**) porosity logs: bulk density, neutron porosity, and helium porosity from core analysis; (**c**) resistivity log; and (**d**) permeability log (using Kozeny's equation) from core analysis [26].

Figure 10 presents a schematic summary of typical well-log responses, illustrating the gas effect in clay-free sandstones (neutron–density crossover) and its absence in greensands. Electrical resistivity logs often produce deceptively low values in glauconitic formations. The high cation-exchange capacity of clay minerals, together with clay-bound and capillary-bound water, enhances the rock's electrical conductivity, resulting in resistivity readings that may underestimate hydrocarbon saturation and complicate petrophysical interpretations.

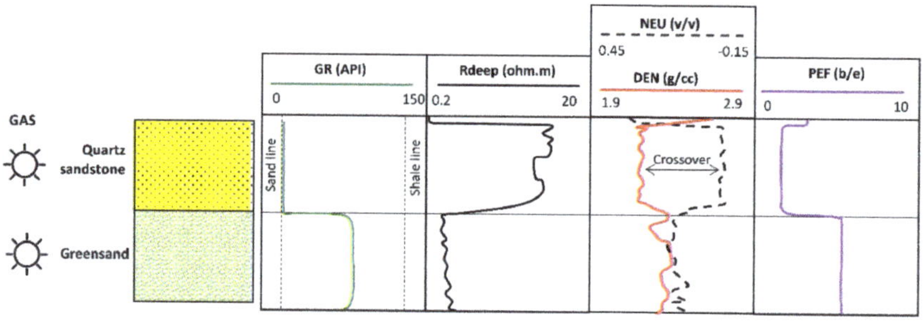

Figure 10. Diagram of conventional well-log responses in quartz sandstone and greensand formations, from left to right: gamma-ray log (GR), deep resistivity log (Rdeep), neutron log (NEU), bulk density log (DEN), and photoelectric factor log (PEF) [32].

Neutron logs tend to overestimate porosity and fluid content due to the significant hydrogen content in iron-rich clay minerals and the associated bound water. Conversely, bulk density logs frequently underestimate porosity, as the higher density of glauconitic and chloritic minerals biases measurements upward. The increased mineral density thus reduces calculated porosity when standard matrix density values are used.

Nuclear magnetic resonance (NMR) logging underestimates mobile fluid volumes because iron-bearing clays increase surface relaxivity, causing rapid decay of NMR signals—seen as shorter T_2 relaxation times. Combined with the dominance of microporosity-hosted fluids, this impedes effective discrimination between movable and bound fluids if conventional T_2 cut-off values are used [1,2,38].

5. Interpretation Framework for Greensand Reservoirs

5.1. Greensands' Petrophysical Controls

The literature identifies four key controls (Figure 11) that underpin porosity, permeability, water saturation, and electrical behaviour in these formations. These controls are interrelated and manifest at both the grain and pore scale, shaping the overall petrophysical signature of greensands. First, the presence of glauconite and associated clay minerals, including chlorite and illite/smectite, exerts a dominant influence. These minerals may be embedded in the framework or occur as pore linings and fillings, directly contributing to heterogeneity in rock texture and anisotropy in physical properties. This mineralogical composition modifies lithological identification, pay zone identification, and permeability characterisation. Their distribution influences grain packing, introduces paramagnetism, and reduces the reliability of standard well-log interpretation techniques.

Second, microporosity—often defined by pore-throat diameters smaller than 1 μm—is a ubiquitous feature in glauconitic grains. These micropores frequently lack connectivity with macropores, reducing effective porosity and flow potential. However, they preferentially retain water over hydrocarbons due to strong capillary forces and water-wet rock conditions. Thus, microporosity impacts reservoir quality, fluid distribution, and S_{wirr} estimations. Third, the specific surface area (SSA) of greensands is high due to the abundance of fine-grained clays and poorly sorted textures. SSA is directly proportional to microporosity and inversely related to permeability and porosity. This relationship challenges upscaling efforts from core to log data and contributes to misleading poro–perm correlations unless mineralogy is explicitly considered.

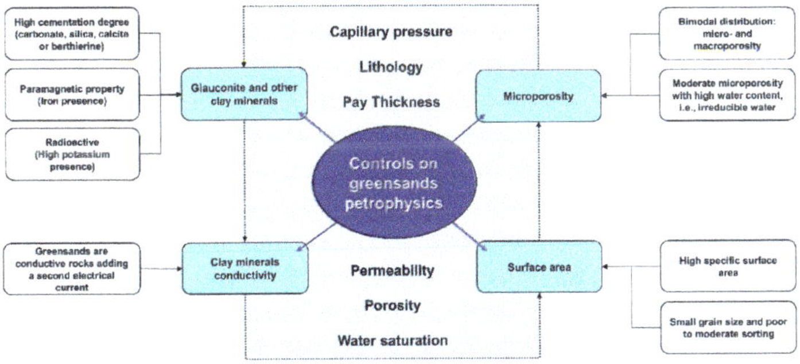

Figure 11. Conceptual framework of petrophysical controls in greensand formations as identified through literature review [32].

Fourth, electrical conductivity induced by clay minerals introduces a secondary current path in resistivity measurements. These conductive clays react electrochemically with formation water, particularly under low-salinity conditions, through cation-exchange mechanisms. The presence of glauconite, which exhibits a moderate to high CEC, increases this conductivity and distorts resistivity-derived saturation calculations. As a result, models based on Archie's law tend to overestimate S_w unless compensated with shale/clay-based models such as Waxman–Smits or Simandoux. These four petrophysical controls—mineralogical heterogeneity, microporosity, high SSA, and clay conductivity—must be accounted for in any reliable interpretation workflow for greensand reservoirs. Neglecting them not only compromises reservoir characterisation but also affects volumetric estimates, net pay quantification, and ultimately development decisions.

5.2. Interpretation Workflows and Tool Recommendations

Interpretation workflows in greensand formations must begin with a careful selection of tools that respond appropriately to their mineralogical and pore-scale complexity. Table 9 outlines these challenges and offers recommended approaches across four categories: (1) lithology identification, (2) porosity estimation, (3) saturation and electrical properties, and (4) permeability determination.

Table 9. Challenges and recommended tools for reservoir property estimation in greensands [32].

Reservoir Property	Challenge	Recommended Tool/Approach (Including Core-Based Methods)
Porosity	Microporosity retention; poor intergranular sorting	NMR (adjusted T_2 cut-offs), brine–helium porosity comparison, SEM image calibration
Permeability	Pore-throat obstruction by glauconite and chlorite; poor sorting	NMR-based models (Timur–Coates, SDR), SSA estimation, core–log regression
Saturation	Overestimated Sw from resistivity; non-Archie behaviour	Waxman–Smits or dual-water models; field-calibrated m and n; CEC measurement
Clay Volume	GR over-response due to high K; mineral misidentification	Spectral gamma-ray, PEF logs, clay typing via mineralogical analysis (XRD, SEM)
Rock Typing	Electrofacies overlap; textural heterogeneity	Multi-mineral log inversion; unsupervised clustering; thin-section validation
Net Pay Determination	Inconsistent cut-offs for Swirr and porosity	NMR calibration, capillary pressure curves, core-calibrated net pay criteria
Reservoir Quality Mapping	Misleading poro–perm trends; variable Swirr	Integration of digital rock physics, electrofacies models, and cross-plots

Table 9. *Cont.*

Reservoir Property	Challenge	Recommended Tool/Approach (Including Core-Based Methods)
Logging Environment	High sensitivity to salinity, temperature, and iron content	In situ environmental corrections, formation-specific interpretation templates

Note: While advanced techniques such as NMR and digital rock analysis are highlighted, core-based methods—including Archimedes (brine saturation) and gas porosimetry—remain valuable for porosity estimation and are often more cost-effective in early-stage reservoir evaluation.

Lithology identification in greensands is complicated by the response of gamma-ray and density–neutron logs to glauconite and other clays. Traditional indicators like spontaneous potential and gamma-ray logs are influenced by high potassium and iron contents, often leading to lithological misclassification. To improve discrimination, tools such as the photoelectric factor (PEF) and spectral gamma-ray logs are recommended. Elemental spectroscopy tools like ECS (Elemental Capture Spectroscopy) offer further refinements by quantifying elemental composition, including iron, silicon, and potassium, which can distinguish glauconitic from other clay-rich zones [71]. Integrating core mineralogy with these logs significantly enhances lithological models. However, defining petrofacies remains challenging due to scattered porosity–permeability trends. Capillary pressure curves and pore-size distribution (PSD) analysis can help clarify pore-throat geometries, and acoustic logs may serve as proxies for reservoir quality in texturally heterogeneous intervals [25].

Porosity estimation is equally complex due to the bimodal distribution and high microporosity in glauconitic sands. Conventional porosity logs—neutron, density, and sonic—must be calibrated against core porosity data, preferably derived from helium expansion or brine saturation methods. NMR tools such as CMR (Combinable Magnetic Resonance) can characterise free-fluid and bound-water volumes and differentiate pore systems. However, the paramagnetic effect of glauconite—due to its iron content—shortens T_2 relaxation times, requiring careful adjustment of cut-off values to avoid underestimating mobile porosity [1,2]. To strengthen interpretation, complementary laboratory techniques such as mercury intrusion porosimetry (MIP) and micro-computed tomography (μCT) are recommended, as they provide independent pore-size distribution data and can help validate NMR-derived porosity estimates. In some cases, μCT may also assist in estimating glauconite content if density contrast permits.

Saturation and electrical properties are strongly affected by clay conductivity and water-wet rock behaviour. Greensands often require modified saturation models, such as Waxman–Smits or Simandoux, since Archie's model typically underestimates Sw due to the additional conductive path introduced by clays. Core-based measurements of cation-exchange capacity (CEC), multi-salinity tests, and capillary pressure curves (e.g., mercury injection capillary pressure, MICP) are vital for calibrating saturation models. These measurements reveal elevated irreducible water saturation and steep capillary pressure profiles, both of which reflect the dominance of micropores in storing formation water [47].

Permeability estimation also demands a combined approach. While NMR-based models such as Timur–Coates or SDR provide log-derived permeability estimates, their accuracy is limited in greensands without calibration. The presence of pore-blocking authigenic clays and cementation reduces pore-throat connectivity, often leading to significant permeability impairment, particularly under low-salinity conditions where clay swelling or migration may occur. Laboratory-determined gas or brine permeability remains essential for ground-truthing log responses. Empirical models, including Kozeny's [57] equation, must be applied cautiously, with a clear understanding of their assumptions and limits.

Finally, geomechanical logging—derived from acoustic and sonic data—can further refine reservoir quality interpretation. These measurements provide information on elastic properties, fracture propagation, and rock brittleness, offering additional insight into stress sensitivity and completion design. In moderately to well-consolidated greensand units, P-impedance and acoustic velocity data can effectively discriminate between productive and non-productive intervals [25].

6. Conclusions and Perspectives

In this review, we synthesised extensive research on the petrophysical controls governing greensand reservoirs, emphasising their distinct mineralogical and pore-scale complexities. Our study confirms that glauconite and associated clay minerals significantly influence porosity, permeability, and water saturation measurements, often leading to challenges in conventional well-log interpretation and reservoir characterisation.

Traditional petrophysical models commonly applied to sandstones frequently fail to accommodate the paramagnetic and conductive effects imparted by glauconite's iron content and high cation-exchange capacity. This commonly results in over- or underestimation of key reservoir parameters, such as porosity and fluid saturations, when standard log interpretations are used without tailored calibrations. Consequently, core analysis remains an indispensable tool for validating log-derived petrophysical properties and developing robust interpretation frameworks.

To improve reservoir evaluation accuracy in greensand contexts, it is important to recognise that glauconite behaves as a conductive, microporous framework mineral with significant magnetic susceptibility and bound water. Integrated workflows are therefore recommended, combining spectral gamma-ray, NMR with provisional adjusted T_2 cut-offs validated against independent measurements, and shale–sand saturation models like Waxman–Smits or Simandoux, calibrated against core and laboratory data from core NMR, MICP, CT, and XRD/mineralogical analysis. These combined approaches help mitigate the misleading effects caused by glauconitic minerals and fine clays, thereby enhancing pay zone identification and net reservoir quality assessments. We also recommend explicit testing of model assumptions and mandatory reporting of key supporting data, including XRD-derived mineral fractions, petrographic modal analysis, bulk magnetic susceptibility, drying protocol, and the rationale for any empirical T_2 or relaxivity adjustments, before applying standard petrophysical models.

Looking forward, future research should prioritise comprehensive core–log–digital rock physics integration to better quantify the impact of mineral heterogeneity and diagenesis on reservoir properties. The development of rock-typing algorithms and machine learning workflows tailored specifically to greensand mineralogy and pore systems holds promise to refine petrophysical models further. Additionally, examining the influence of geochemical and environmental factors on glauconite stability and reservoir quality will support improved predictive capabilities under variable production conditions.

Given the increasing exploration interest in greensand reservoirs worldwide, greater collaboration between academia, industry practitioners, and technology innovators is essential to evolve interpretation standards and deploy best practices. By embracing the unique petrophysical nature of greensands and advancing specialised tools and workflows, the petroleum sector can enhance hydrocarbon recovery efficiencies and optimise reservoir management in these complex lithologies.

Author Contributions: Conceptualisation, D.N.-P.; methodology, D.N.-P.; data curation, D.N.-P.; writing—original draft preparation, D.N.-P.; writing—review and editing, Q.F., P.L., A.V.A. and J.V.P.; visualisation, D.N.-P.; funding acquisition, D.N.-P. All authors have read and agreed to the published version of the manuscript.

Funding: This research work was funded by the Agencia Nacional de Investigación y Desarrollo de Chile (ANID) and the scholarship programme Doctorado en el Extranjero Becas Chile (Grant No. 2018-72190176). The APC was funded by the Facultad de Ingeniería, Universidad de Magallanes, Chile.

Acknowledgments: This work is derived from the first author's doctoral research conducted at the University of Leeds. The author, D.N., acknowledges the support provided by Facultad de Ingeniería of the Universidad de Magallanes.

Conflicts of Interest: Authors Aníbal Velásquez Arauna was employed by the company ENAP Magallanes and Jose Valderrama Puerto was employed by the company ENAP SIPETROL S.A. The remaining authors declare that the research was conducted in the absence of any commercial or financial relationships that could be construed as a potential conflict of interest.

References

1. Dodge, W.S.; Shafer, J.L.; Klimentidis, R.E. Capillary pressure: The key to producible porosity. In Proceedings of the 37th SPWLA Annual Logging Symposium, New Orleans, LA, USA, 16–19 June 1996; p. SPWLA-1996-J.
2. Rueslåtten, H.; Eidsemo, T.; Slot-Petersen, C. NMR studies of iron-rich sandstone oil reservoir. In Proceedings of the International Symposium of Society of Core Analysts, The Hague, The Netherlands, 14–16 September 1998; p. SCA-9821.
3. Durand, C.; Cerepi, A.; Brosse, E. Effect of pore-lining chlorite on petrophysical properties of low-resistivity sandstone reservoir. In Proceedings of the 2000 SPE Annual Technical Conference and Exhibition, Dallas, TX, USA, 1–4 October 2000; p. SPE-63070. [CrossRef]
4. Bansal, U.; Banerjee, S.; Nagendra, R. Is the rarity of glauconite in Precambrian Bhuima Basin in India related to its chloritization? *Precambrian Res.* **2020**, *336*, 105509. [CrossRef]
5. Thompson, G.R.; Hower, J. The mineralogy of glauconite. *Clays Clay Miner.* **1975**, *23*, 289–300. [CrossRef]
6. Odin, G.S.; Letolle, R. Glauconitisation and phosphatization environments: A tentative comparison. *SEPM Spec. Publ.* **1980**, *29*, 227–237.
7. Harder, H. Boron content of sediments as a tool in facies analysis. *Sediment. Geol.* **1969**, *4*, 153–175. [CrossRef]
8. Patchett, J.G.; Wiley, R.; El Bahr, M. Modeling the effects of glauconite on some openhole logs from the lower Senonian in Egypt. In Proceedings of the SPWLA 34th Annual Logging Symposium, Calgary, AB, Canada, 13–16 June 1993; p. SWPLA-1993-RR.
9. Archie, G.E. The electrical resistivity log as an aid in determining some reservoir characteristics. *Trans. AIME* **1942**, *146*, 54–62. [CrossRef]
10. Worthington, P.F. The evolution of shaly-sand concepts in reservoir evaluation. *Log Anal.* **1985**, *26*, 23–40.
11. Waxman, M.H.; Smits, L.J.M. Electrical conductivities in oil-bearing shaly sands. *Soc. Pet. Eng. J.* **1968**, *8*, 107–122. [CrossRef]
12. Clavier, C.; Coates, G.; Dumanoir, J. Theoretical and experimental bases for the dual-water model for interpretation of shaly sands. *Soc. Pet. Eng. J.* **1984**, *24*, 153–167. [CrossRef]
13. Worthington, P.F. Net pay—What is it? What does it do? How do we quantify it? How do we use it? In Proceedings of the SPE Asia Pacific Oil and Gas Conference & Exhibition. Jakarta, Indonesia, 4–6 August 2009; p. SPE-123561. [CrossRef]
14. Hatcher, G.B.; Chen, H.; Rahman, S.S.; Hogg, P.F. Evaluating formation damage risks in a glauconitic sandstone reservoir: A case history from the offshore North West Shelef of Australia. In Proceedings of the SPE Asia Pacific Oil and Gas Conference, Adelaide, Australia, 28–31 October 1996; p. SPE-37014. [CrossRef]
15. Zhang, Y.; Salisch, H.A.; Arns, C. Permeability evaluation in a glauconite-rich formation in the Carnarvon Basin, Western Australia. *Geophysics* **2000**, *65*, 46–53. [CrossRef]
16. Thomas, W.H.; Ringen, J.K.; Rasch, S.O. Effect of glauconite on petrophysical properties as revealed by core analysis. In Proceedings of the International Symposium of the Society of Core Analysts, Pau, France, 21–24 September 2003; p. SCA2003-32.
17. Klein, J.D.; Little, L.; Scheihing, M.; Seifert, D. Formation evaluation and permeability prediction in a highly heterogeneous reservoir: The Kuparuk C-Sand. In Proceedings of the SWPLA 47th Annual Logging Symposium, Veracruz, Mexico, 4–7 June 2006; p. SPWLA-2006-FFF.
18. Hossain, Z. Rock-Physics Modelling of the North Sea Greensand. Ph.D. Thesis, Technical University of Denmark, Kongens Lyngby, Denmark, 2011.
19. Prayoga, O.A.; Wicaksono, B.; Setyoko, S.; Wibowo, A.; Wijaksono, E.; Sulistyono; Momem, M.; Zulmi, I. Integrated analysis for reservoir characterization of low resistivity glauconitic shaly sand reservoir of Miocene Ngryong Sandstone: Implications for saturation modelling at low contrast hydrocarbon pay zone. In Proceedings of the 2nd SPWLA Asia Pacific Technical Symposium, Bogor, Indonesia, 26–27 November 2018; p. SPWLA-2018-1810.
20. Nong, K.; Chen, S.; Ren, Z.; Zeng, M. Analysis on glauconite research trends based on citespace knowledge graph. *Minerals* **2024**, *14*, 1260. [CrossRef]

21. Dasi, E.; Rudmin, M.; Banerjee, S. Glauconite applications in agriculture: A review of recent advances. *Appl. Clay Sci.* **2024**, *253*, 107368. [CrossRef]
22. Broger, K.; Syhlonyk, G. Glauconite sandstone exploration: A case study from the Lake Newell project, Southern Alberta, Canada. In Proceedings of the AAPG Annual Convention, Denver, CO, USA, 12–15 June 1994.
23. Taber, D.R.; Vickers, M.K.; Winn, R.D., Jr. The definition of the Albian 'A' Sand reservoir fairway and aspects of associated gas accumulation in the North Celtic Sea Basin. *Geol. Soc. Lond. Spec. Publ.* **1995**, *93*, 227–244. [CrossRef]
24. Garamendi, R.; Atau, H. Estrategia de explotacion en reservorios de bajo restablecimiento de presion. In Proceedings of the INGEPET 1999, Lima, Peru, 26–29 October 1999.
25. Diaz, E.; Prasad, M.; Mavko, G.; Dvorkin, J. Effect of glauconite on the elastic properties, porosity and permeability of reservoir rocks. *Lead. Edge* **2003**, *22*, 42–45. [CrossRef]
26. Hossain, Z.; Grattoni, C.A.; Solymar, M.; Fabricius, I.L. Petrophysical properties of greensand as predicted from NMR measurements. *Pet. Geosci.* **2011**, *17*, 111–125. [CrossRef]
27. Yang, X.F.; Ma, Z.Z.; Zhou, Y.B.; Zhang, Z.W.; Liu, Y.M.; Wang, D.D.; Zhao, Y.B. Reservoir characteristics and hydrocarbon accumulation of the glauconitic sandstone in the Tarapoa Block, Oriente Basin, Ecuador. *J. Pet. Sci. Eng.* **2019**, *173*, 558–568. [CrossRef]
28. López-Quirós, A.; Escutia, C.; Sánchez-Navas, A.; Nieto, F.; Garcia-Casco, A.; Martín-Algarra, A.; Evangelinos, D.; Salabarnada, A. Glaucony authigenesis, maturity and alteration in the Weddell Sea: An indicator of paleoenvironmental conditions before the onset of Antarctic glaciation. *Sci. Rep.* **2019**, *9*, 13580. [CrossRef]
29. Odin, G.S.; Matter, A. De glauconiarum origin. *Sedimentology* **1981**, *28*, 611–641. [CrossRef]
30. Baldermann, A.; Warr, L.N.; Grathoff, G.H.; Dietzel, M. The rate and mechanism of deep-sea glauconite formation at the Ivory Coast-Ghana marginal ridge. *Clays Clay Miner.* **2013**, *61*, 258–276. [CrossRef]
31. López-Quirós, A.; Sánchez-Navas, A.; Nieto, F.; Escutia, C. New insights into the nature of glauconite. *Am. Mineral.* **2020**, *105*, 674–686. [CrossRef]
32. Navarro-Perez, D.J. Improved Reservoir Characterisation of a Chilean Tight Sandstone Reservoir. Ph.D. Thesis, University of Leeds, Leeds, UK, 2024.
33. Wilson, L.; Wilson, M.J.; Green, J.; Patey, I. The influence of clay mineralogy on formation damage in North Sea reservoir sandstone: A review with illustrative examples. *Earth-Sci. Rev.* **2014**, *134*, 70–80. [CrossRef]
34. Hossain, Z.; Fabricius, I.L.; Christense, H.F. Elastic and nonelastic deformation of greensand. *Lead. Edge* **2009**, *28*, 86–88. [CrossRef]
35. Atahualpa, G. Efecto de la Glauconita en las Propiedades Petrofísicas del Reservorio Arenisca "T" Superior del Bloque Tarapoa. Bachelor's Thesis, Universidad Central de Ecuador, Quito, Ecuador, 2013.
36. Guanochanga, J. Modelo Geológico, Caracterización Petrofísica y Cálculo de Reservas, en la Arenisca "T" Superior del Campo Mariann 4A y Mariann Norte en el Bloque Tarapoa. Bachelor's Thesis, Universidad Central de Ecuador, Quito, Ecuador, 2013.
37. Pittman, E.D. Porosity, diagenesis and productive capability of sandstone reservoirs. *Soc. Econ. Paleontol. Mineral. (SEPM) Spec. Publ.* **1979**, *26*, 159–173.
38. Slot-Petersen, C.; Eidesmo, T.; White, J.; Rueslatten, H.G. NMR formation evaluation applications in a complex low-resistivity hydrocarbon reservoir. In Proceedings of the SPWLA 39th Annual Logging Symposium, Keystone, CO, USA, 26–29 May 1998; p. SPWLA-1998-TT.
39. Markley, M.; Seminario, F.; Gabulle, J.; Luquez, J. Microporosity and laminations in Non-Archie reservoirs create challenges for water-saturation computation and reserves evaluation: Camisea, Peru. In Proceedings of the SPE Latin American & Caribbean Petroleum Engineering Conference, Lima, Peru, 1–3 December 2010; p. SPE-139424. [CrossRef]
40. Hossain, Z.; Mukerji, T.; Dvorkin, J.; Fabricius, I.L. Rock physics model of glauconite greensand from the North Sea. *Geophysics* **2011**, *76*, E199–E209. [CrossRef]
41. Hugget, J.M. Glauconites. In *Encyclopedia of Geology*; Alderton, D., Scott, A.E., Eds.; Academy Press: Cambridge, MA, USA, 2021; pp. 334–340.
42. Burst, J.F. Mineral heterogeneity in "glauconite pellets". *Am. Mineral. J. Earth Planet. Mater.* **1958**, *43*, 481–497.
43. Hower, J. Some factors concerning the nature and origin of glauconite. *Am. Mineral. J. Earth Planet. Mater.* **1961**, *46*, 313–334.
44. Cimbalnikova, A. Chemical variability and structural heterogeneity of glauconites. *Am. Mineral. J. Earth Planet. Mater.* **1971**, *56*, 1385–1392.
45. Cimbalnikova, A. Influence of 10Å/14Å interlayering on the layer charge of glauconites. *Am. Mineral. J. Earth Planet. Mater.* **1971**, *56*, 1393–1398.
46. Abudelgawad, G.; Page, A.L.; Lund, L.J. Chemical weathering of glauconite. *Soil Sci. Soc. Am. J.* **1975**, *39*, 567–571. [CrossRef]
47. El-Amamy, M.M.; Page, A.L.; Abudelgawad, G. Chemical and mineralogical properties of glauconitic soils as related to potassium depletion. *Soil Sci. Soc. Am. J.* **1982**, *46*, 426–430. [CrossRef]
48. McPhee, C.; Reed, J.; Zubizarreta, I. *Core Analysis: A Best Practice Guide, 64*; Elsevier: Amsterdam, The Netherlands, 2015.

49. Navarro-Perez, D. Quantifying the porosity of natural reservoirs with nuclear magnetic resonance. *Nat. Rev. Earth Environ.* **2023**, *4*, 599. [CrossRef]
50. Coates, G.R.; Xiao, L.; Prammer, M.G. *NMR Logging Principles and Applications*; Elsevier Science: Houston, TX, USA, 1999.
51. Hossain, Z.; Zhou, Y. Petrophysics and rock physics modeling of diagenetically altered sandstone. *Interpretation* **2015**, *3*, 107–120. [CrossRef]
52. Navarro-Perez, D.; Fisher, Q.; Allshorn, S.; Grattoni, C.; Lorinczi, P. Multi-salinity core flooding study in clay-bearing sandstones, a contribution to geothermal reservoir characterisation. *Adv. Geosci.* **2024**, *62*, 71–80. [CrossRef]
53. Van Olphen, H.; Fripiat, J.J. Data handbook for clay materials and other non-metallic minerals. *Soil Sci.* **1979**, *131*, 62. [CrossRef]
54. Timur, A. An investigation of permeability, porosity, and residual water saturation relationships. In Proceedings of the SPWLA 9th Annual Logging Symposium, New Orleans, LA, USA, 23–26 June 1968; p. SPWLA-1968-J.
55. Kenyon, W.E. Petrophysical principals of applications of NMR logging. *Log ASnal.* **1997**, *38*, 21–43.
56. Kozeny, J. Ueber kapillare Litung des Wassers im Boden. *Sitz.-Berichte Der Kais. Akad. Der Wiss. Wien* **1927**, *136*, 271–306.
57. Hossain, Z.; Cohen, A.J. Relationship among porosity, permeaility, electrical and elastic properties. In Proceedings of the 2012 SEG Annual Meeting, Las Vegas, NV, USA, 4–9 November 2012; p. SEG-2012-1496.
58. Zhang, Y.J.; Lollback, P.A.; Rojahn, J.S.; Salisch, H.A.; Stuart, W.J. A methodology for estimating permeability from well logs in a formation of complex lithology. In Proceedings of the SPE Asia Pacific Oil and Gas Conference and Exhibition, Adelaide, Australia, 28–31 October 1996; p. SPE-37025-MS. [CrossRef]
59. Simpson, G.A.; Fishman, N.S.; Hari-Roy, S. New Nuclear Magnetic Resonance Log T2 Cut-off Interpret Parameters for the Unconventional Tight Oil of the Bakken Petroleum System Using 2-D NMR Core Laboratory Measurements on Native State and Post-Cleaned Core Samples. In Proceedings of the SPWLA 59th Annual Logging Symposium, London, UK, 2–6 June 2018; p. SPWLA-2018-GGGG.
60. Cluff, R.M.; Byrnes, A.P. Relative permeability in tight gas sandstone reservoirs—The "permeability jail" model. In Proceedings of the SPWLA 51st Annual Logging Symposium, Perth, Australia, 19–23 June 2010; p. SPWLA-20120-58470.
61. Weaver, C.E. *Clays, Muds, and Shales*; Elsevier: Amsterdam, The Netherlands, 1989.
62. Pratama, E.; Ismail, M.S.; Ridha, S. An integrated workflow to characterize and evaluate low resistivity pay and its phenomenon in a sandstone reservoir. *J. Geophys. Eng.* **2017**, *14*, 513–519. [CrossRef]
63. Worthington, P.F. Recognition and evaluation of low-resistivity pay. *Pet. Geosci.* **2000**, *6*, 77–92. [CrossRef]
64. Sneider, R.M. Worldwide examples of low resistivity pay. *Houst. Geol. Soc. Bull.* **2003**, *45*, 47–59.
65. Bardon, C.; Pied, B. Formation water saturation in shaly sands. In Proceedings of the SWPLA 10th Annual Logging Symposium, Houston, TX, USA, 25–28 May 1969; p. SWPLA-1969-Z.
66. Hossain, Z.; Mukerji, T.; Fabricius, I.L. Vp-Vs relationship and amplitude variation with offset modeling of glauconitic greensand. *Geophys. Prospect.* **2012**, *60*, 117–137. [CrossRef]
67. Gonzalez-Gonzalez, A.; Valderrama, J.M.; Gschaider, C.T.; Carcamo, R.A.; Verdugo-Dobronic, M.; Canessa, N.D.; Perez-Perez, A.; Velasquez-Arauna, A.; Sanchez-Ojeda, J.; Ahumada-Villar, M.; et al. Desarrollo del play tight gas zona glauconitica, Bloque Arenal, Cuenca de Magallanes—Chile: Caso estudio. In Proceedings of the 10° Congreso de Exploracion y Desarrollo de Hidrocarburos, Mendoza, Argentina, 5–9 November 2018.
68. Aimar, E.; Cevallos, M.; Cangini, A.; Mas-Cattapan, F.; Vega, V. Extension y desarrollo de los reservorios de baja permeabilidad del yacimiento Campo Indio, Formación Magallanes (Maastricchtiano Tardío-Daniano), Cuenca Austral Argentina. In Proceedings of the 10° Congreso de Exploracion y Desarrollo de Hidrocarburos, Mendoza, Argentina, 5–9 November 2018.
69. Kennedy, W.D.; Herrick, D.C. Conductivity models for Archie rocks. *Geophysics* **2012**, *77*, WA109–WA128. [CrossRef]
70. Eslinger, E.; Pevear, D. *Clay Minerals for Petroleum Geologists and Engineers, SEPM Short Course 22*; Society for Sedumentary Geology: Tulsa, OK, USA, 1988.
71. Schlumberger. ECS Elemental Capture Spectroscopy Sonde. 2006. Available online: https://www.slb.com/-/media/files/fe/brochure/ecs-brochure (accessed on 26 February 2021).

minerals

MDPI

Review

Advanced Trends in Shale Mechanical Inhibitors for Enhanced Wellbore Stability in Water-Based Drilling Fluids

Erxin Ai [1], Qi Li [1,*], Zhikun Liu [1], Bo Zhang [2], Liupeng Wang [1] and Chengyun Ma [3]

1 College of Petroleum Engineering, Xi'an Shiyou University, Xi'an 710065, China; aex@xsyu.edu.cn (E.A.); lzk12431@xsyu.edu.cn (Z.L.); wangliupeng@xsyu.edu.cn (L.W.)
2 Explorement & Development Project Department of Mahu Area of Xinjiang Oilfield Company, Karamay 834000, China; petroup24@163.com
3 Mechanical Engineering College, Xi'an Shiyou University, Xi'an 710065, China; mcy0000@163.com
* Correspondence: liqi@xsyu.edu.cn

Abstract: Wellbore instability, particularly in shale formations, presents a great challenge to modern drilling operations. Although conventional chemical inhibitors are frequently employed in water-based drilling fluids, they may not always function in highly reactive or naturally fractured shale formations. In recent years, mechanical inhibitors have attracted attention as a complementary solution to chemical methods. These inhibitors, which include carbon-based, silicon-based, metal-based, and mineral-based particle materials, provide structural support to the wellbore by physically plugging fractures and sealing microfractures. This paper presents a comprehensive review of the mineral types associated with shale wellbore instability as well as a critical analysis of the mechanisms, categories, and effectiveness of mechanical inhibitors in enhancing wellbore stability. The development challenges and prospects of mechanical inhibitors in water-based drilling fluids are also discussed. This review emphasizes the potential of mechanical inhibitors in reducing fluid invasion, preventing wellbore collapse, and improving overall drilling efficiency, which will help facilitate the development and large-scale application of mechanical inhibitors in drilling fluids.

Keywords: drilling fluid; shale inhibitors; wellbore stability; clay mineral; physical plugging

1. Introduction

The growing demand for oil and gas consumption has driven the need to drill deeper, more complex, and geologically challenging wells [1]. The drilling operation is the first step in oil and gas extraction. Drilling wells is one of the most critical activities in producing oil and gas from reservoirs. However, as drilling operations venture into unconventional reservoirs and extreme environments such as deepwater and high-pressure high-temperature (HPHT) conditions, significant engineering challenges arise, among which wellbore instability is one of the most critical issues [2,3]. The causes of wellbore instability can be divided into three main categories: mechanical factors caused by in situ stresses, erosion factors caused by fluid circulation, and chemical factors caused by the interaction of fluids with the formation [2,4–6]. The instability of the wellbore can lead to various problems, including stuck pipe incidents, wellbore collapse, lost circulation, and difficulties in running casing. These issues not only increase the non-productive time (NPT) but also significantly escalate drilling costs. It is reported that wellbore instability has resulted in millions of man-hours of NPT and economic losses of up to $6 billion per year [7,8]. Beyond the economic impact, wellbore instability also poses safety risks to personnel and equipment, and is, therefore, a critical issue that must be addressed in drilling operations.

Shale wellbore instability has long posed a major challenge in drilling operations, particularly in unconventional oil and gas reservoirs. Wellbore instability is considered to be an undesirable condition in which the open hole interval cannot maintain its gauge size, shape, and structural integrity [9]. The main reason for wellbore instability in shale formations is that free water and ions in the water-based drilling fluids (WBDFs) penetrate into the formation and react with the shale, thereby reducing the strength of the shale rock. Shale formations, which constitute a significant portion of the world's sedimentary basins, are characterized by their complex mineral composition, high sensitivity, and reactivity [10,11]. Shales are highly heterogeneous, containing a mixture of minerals like clays (smectite, illite, and kaolinite), quartz, feldspar, and organic matter [12]. Clay minerals are especially critical to wellbore instability due to their strong water affinity and tendency to swell when exposed to aqueous fluids. Moreover, shales also tend to exhibit anisotropy, with bedding planes being highly sensitive to stress changes or fluid intrusion [13]. The combination of clay swelling and stress anisotropy can lead to localized failure along these planes, contributing to wellbore collapse. In some cases, the situation is further exacerbated by chemical interactions between the drilling fluid and the shale, such as ion exchange processes, which can weaken the shale structure [14–16].

The alteration of shale mechanical and physical properties in WBDFs can lead to wellbore instability, which manifests as sloughing, caving, or hole enlargement. Some drilling practices have proven that oil-based drilling fluids (OBDFs) are outstanding in inhibiting shale hydration and preventing wellbore instability, but they are associated with environmental risks, including contamination of surface and groundwater resources, as well as challenges in treating and disposing of oil-contaminated drill cuttings [17,18]. The environmental friendliness and low cost of WBDFs are their greatest advantages over OBDFs, but how to improve the inhibition ability of WBDFs in reactive shale formations is the key to their widespread application. The current prevalent methods for improving the inhibitory properties of WBDFs focus on the addition of a wide variety of chemical inhibitors to the drilling fluid, including salts, polymers, ionic liquids, and deep eutectic solvents [19–23]. These inhibitors function by reducing water activity, stabilizing clay surfaces, or preventing ion exchange processes. Many researchers have reviewed the synthesis methods and categories of chemical inhibitors, as well as their inhibition mechanisms, application effects, and prospects in WBDFs [24–28]. Chemical inhibitors have been proven to be effective in many cases, but their performance may be limited in certain geological conditions, particularly in formations with highly reactive clays or where the formation is naturally fractured [29].

In recent years, the use of mechanical inhibitors has received increasing attention as a complementary method to chemical inhibitors. The advantage of mechanical inhibitors is that they can provide structural support to the wellbore, thereby reducing the possibility of collapse or other instability-related problems. These inhibitors prevent drilling fluids from penetrating the formation by bridging fractures and sealing microfractures [30,31]. Mechanical inhibitors are particularly effective in formations where chemical methods alone may not be sufficient, such as in naturally fragile formations [32]. Mechanical inhibitors in WBDFs are typically solid-phase additives that act by forming a physical barrier between the pore fracture and the surrounding formation, thereby reducing the interaction between the drilling fluid and the formation rock. Unlike chemical inhibitors, mechanical inhibitors work primarily through physical plugging, including the formation of a protective mud cake, particle bridging, and the strengthening of the wellbore through mechanical reinforcement [33,34].

The purpose of this review is to provide a comprehensive overview of the current state of knowledge regarding the use of mechanical inhibitors in WBDFs to improve wellbore stability. It will examine the mechanisms of these inhibitors, the types currently available, and their effectiveness in various geological environments. Although existing reviews have provided a comprehensive summary of how to improve the inhibitory properties of WBDFs, much of the current literature focuses on the utilization of shale chemical inhibitors to improve wellbore stability, and there is a lack of reviews on shale mechanical inhibitors. Mechanical inhibitors have shown unparalleled advantages in inhibiting highly reactive shale formations or fractured formations. The objective of this review is to highlight the potential of mechanical inhibitors as a key tool in the ongoing effort to improve wellbore stability in modern drilling operations. This review is intended to promote the use of mechanical inhibitors and to emphasize the inhibitory capabilities of mechanical inhibitors in order to achieve more efficient, safer, and cost-effective drilling practices in the future.

In this work, the application of mechanical inhibitors in WBDFs in the past five years is summarized, focusing on their inhibition mechanisms, classification, and effectiveness. The first part following the introduction summarizes the properties of clay minerals related to wellbore instability in detail. The second section describes the mechanism of mechanical inhibitors and provides a comprehensive classification and summary. Finally, the prospects and challenges of using shale mechanical inhibitors in WBDFs are emphasized before the conclusion.

2. How Clay Minerals Are Related to Wellbore Instability

Clay minerals perform a critical function in wellbore instability, particularly in shale-rich formations. Their unique mineral structure, swelling tendencies, and interactions with WBDFs render them one of the most challenging factors in maintaining wellbore stability. Shale formations typically contain a high percentage of clay minerals, which exhibit unpredictable behavior due to the clay minerals' sensitivity to changes in pressure, temperature, and fluid chemistry [35,36]. When clay minerals absorb water or react with ions in the drilling fluid, they swell and weaken the surrounding rock matrix, ultimately compromising the structural integrity of the wellbore [37,38]. Therefore, it is vital to understand the mineralogical composition and hydration behavior of clays for the design of drilling fluids and wellbore stabilization strategies.

Clay minerals in shale mainly include montmorillonite (MMT), illite, kaolinite, chlorite, and illite/montmorillonite mixed layers. The clay minerals are phyllosilicates in which the individual layers are composed primarily of tetrahedral and octahedral sheets [39]. However, the composition, molecular structure, and chemical properties of different clay minerals vary greatly. The interaction between shale and drilling fluids is highly dependent on the type and amount of clay minerals in the shale [40]. The main characteristics of clay minerals in shale are summarized in Table 1.

Table 1. The characteristic parameters of clay minerals in shale.

Clay Minerals	Layer Arrangement	Chemical Elements	CEC (meq/100 g)	Basal Spacing (nm)	Ref.
Montmorillonite	2:1	$(Na,Ca)_{0.3}(Al,Mg)_2Si_4O_{10}(OH)_{2n}(H_2O)$	80–150	1.2	[41]
Illite	2:1	$K_{1-1.5}Al_4[Si_{7-6.5}Al_{1-1.5}O_{20}](OH)_4$	10–30	1.0	[42]
Kaolinite	1:1	$Al_4[Si_4O_{10}](OH)_8$	3–15	0.72	[43]
Chlorite	2:1	$Mg_5Al_2Si_3O_{10}(OH)_8$	10–40	1.4	[44]

2.1. *Smectite*

Smectite is a group of swellable 2:1 phyllosilicate clay minerals that possesses exceptional hydration swelling capacity, high specific surface area, and high cation exchange capacity [45]. MMT is a typical mineral of the smectite group and is often referred to as bentonite in the industry. The unit cell structure of MMT is a typical 2:1 layered-clay mineral structure, which consists of two layers of two silica tetrahedral sheets (T) and a layer of alumina octahedral sheet (O) (Figure 1). MMT layers retain a platelet structure and they assemble themselves together by stacking up each of the clay layers in a recurrent manner, forming a layered structure with a regular gap within consecutive layers [46]. Due to isomorphic substitution, Si^{4+} is substituted by Al^{3+} in the tetrahedral sheet, and Al^{3+} is replaced by Mg^{2+} in the octahedral layer, resulting in the negative charge of the MMT layer. Consequently, some exchangeable cations, such as Na^+ and Ca^{2+}, exist in the interlayer space to neutralize the negative charges of the clay layers [47]. The thickness of a single layer of MMT is 0.96 nm, but the diameter varies from nanometers to micrometers [48].

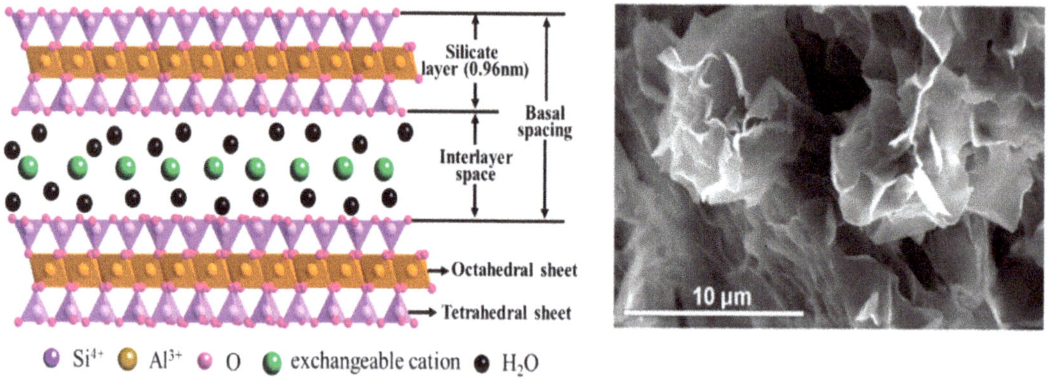

Figure 1. The crystal structure of MMT and corresponding SEM image.

The main characteristic of MMT is the existence of wide gaps between clay platelets, which results in weak electrostatic forces [23]. Additionally, MMT is naturally hydrophilic due to the presence of hydrated exchangeable cations. When water molecules are adsorbed on the interlayers, the smectite clay minerals can swell, thereby increasing the basal spacing. In fact, there is a clear positive correlation between the interlayer spacing of MMT and its degree of hydration: by increasing the amount of water molecules, the crystalline lattice expands [49]. The swelling of smectite in an aqueous solution is also known as hydration. Hydration is the movement or separation of clay layers, especially those in a parallel arrangement, resulting in an increase in the interlayer space or the expansion of the planes of layers (Figure 2). The hydration of MMT causes it to expand beyond its original volume, manifesting as wellbore closure or collapse during drilling and leading to wellbore instability [50–53].

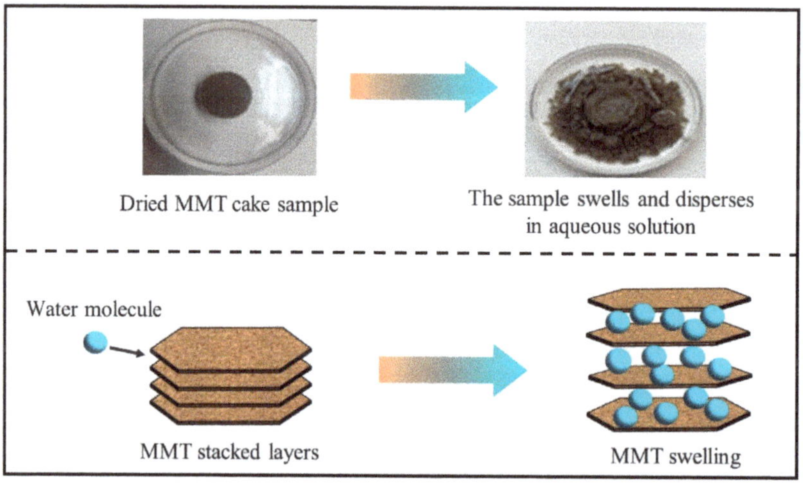

Figure 2. Experimental results and schematic diagram of MMT swelling and dispersion in aqueous solutions.

2.2. Illite

Similar to the crystal structure of MMT, illite is also a 2:1 type clay mineral, that is, the illite mineral structure contains two silica tetrahedrons and one alumina octahedron. However, unlike MMT, which is easy to hydrate and swell, illite has less potential for hydration and swelling in aqueous solutions. Illite in the formation has a variety of morphologies, from undulose platelets ("cornflake" texture) at one end of the spectrum to laths, fibers, wisps, or ribbons ("hairy" illite) at the other [54] (Figure 3). Illite is produced by the alteration of muscovite and feldspar in extreme weathering and hydrothermal environments. This chemical process causes the interlayer space between individual clay crystals to be occupied by weakly hydrated potassium ions (K^+), which further inhibits expansion [55]. Alternatively, smectite can also be transformed at HTHP into an illite. The obtained illite–smectite mixed layer may be more prone to swell than its original version [56]. The lack of expanding lattices makes it difficult for water to penetrate the interlayer space of illite, resulting in a weak swelling capacity. This strong interlayer bonding may be related to the higher charge near the surface of the tetrahedral lamellae. For illite, cation exchange can still occur at the exterior surfaces of each layer, but the volume increase is much smaller than that caused by the hydration of MMT.

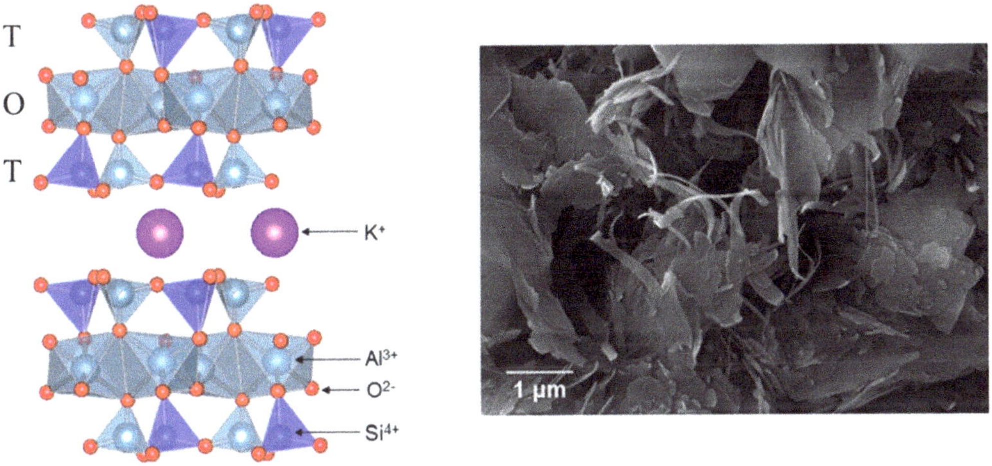

Figure 3. The crystal structure of illite and corresponding SEM image [57].

2.3. Kaolinite and Chlorite

Kaolinite and chlorite are two other members of the clay mineral family that are considered non-swelling clay minerals, although they have distinct crystal structures. Kaolinite has a 1:1 layered structure, consisting of one tetrahedral sheet connected to one octahedral sheet. It features Al^{3+} in the octahedral sites and Si^{4+} in the tetrahedral sites [58]. The hydrogen bonding in kaolinite occurs between oxygen atoms and hydroxyl ions of the paired layers, fixing its structure and resulting in no expansion or a low swelling tendency when wetted [59]. Despite being recognized as a non-expansive clay mineral, kaolinite can be dispersed in aqueous solutions. Consequently, shales with high kaolinite content often exhibit brittle behavior, posing challenges to the mechanical stability of wellbores [56,60]. There is considerable evidence that shales with a significant amount of kaolinite are more susceptible to deposition when exposed to WBDFs [61]. Therefore, when drilling shale formations containing kaolinite, it is necessary not only to consider inhibiting the dispersion of kaolinite, but also to improve the ability of WBDFs to carry cuttings so that they can be prevented from being deposited.

Chlorites are hydrous aluminosilicates characterized by a 2:1 structural arrangement, wherein an alumina octahedron is sandwiched between two tetrahedral sheets of silica within the interlayers. They are typically rich in iron and magnesium and have a wide range of specific compositions, resulting in significant differences in their physical and chemical properties [62,63]. Petroleum geologists found that deeply buried sandstones containing chlorite coatings have exceptionally high porosity and permeability compared to non-chlorite-coated sandstones [64,65]. These highly penetrative formations facilitate the migration of oil and gas. However, as a non-swelling clay mineral, chlorite has significantly less hydration and swelling capacity than smectite minerals, and, therefore, few reports have been recorded on its effects on wellbore integrity.

2.4. *Illite/Smectite Mixed Layer*

Although the hydration of smectite has traditionally been considered the primary cause of wellbore instability, extensive core mineral composition analyses revealed that the pure smectite content in formations is minimal. Instead, the illite/smectite mixed layer emerges as the predominant mineral component of the core. Shale samples from the Ordos Basin in China demonstrated that the illite/smectite mixed layer accounts for 36% to 97% of the clay content, and the illite is mainly found in the range of 21%~32%. Despite the absence of highly swelling smectite in the clay, the presence of the illite/smectite mixed layer obviously makes the shale display water sensitivity [66]. The illite/smectite mixed layer is gradually transformed from the smectite mineral through a series of intermediate steps. It represents a transitional mineral between illite and smectite, and is composed of illite and smectite crystalline layers along the C-axis. Therefore, the hydration capacity of the illite/smectite mixed layer should be between that of illite and smectite [67]. The presence of mixed-layer illite/smectite can significantly influence the mechanical properties of the shale, as the mixed-layer structure imparts unique swelling and hydration characteristics that differ from those of pure smectite or illite.

3. Wellbore Instability Caused by Clay Minerals

The patterns of wellbore instability occurring in shale formations primarily involve wellbore collapse, wellbore enlargement, and wellbore closure. These problems are mainly caused by the unique mineralogical and geomechanical properties of shale, particularly the presence of various clay minerals [68]. The presence of different clay minerals significantly influences the mechanical properties of the shale and its behavior under drilling conditions. For instance, MMT, the most problematic clay mineral, is particularly notorious for its high swelling capacity. When exposed to water, it can swell dramatically, increasing the risk of wellbore instability. This swelling leads to a reduction in effective stress acting on the wellbore, causing the walls to become more susceptible to collapse [69]. Moreover, the resulting excess pore pressure can further exacerbate instability, creating a cycle of failure that can halt drilling operations and lead to significant financial losses [70]. Illite is relatively stable compared to MMT but can become brittle when dehydrated. This brittleness can result in the formation of fractures that compromise wellbore stability under certain stress conditions [71].

Kaolinite, in contrast, is characterized by its non-expandable nature and lower plasticity, which can improve wellbore stability. However, in high-stress conditions, it may still fail, particularly in formations with a complex mineralogical composition [14]. Chlorite can contribute positively to stability, but its behavior can vary depending on the geological context [72]. In some cases, the presence of chlorite can enhance the mechanical properties of the shale, while in others, it can lead to localized instability, particularly when combined with other clay minerals that alter the stress distribution in the formation [73]. The complex interaction between the drilling fluid and the mineralogical composition of shale formations plays a vital role in determining wellbore stability. A comprehensive understanding of the mineralogical characteristics and the geomechanical behavior of shale is essential for developing effective drilling strategies that ensure successful wellbore stability.

4. Mechanism of Mechanical Inhibitors

During the drilling process, the overbalance between the drilling fluid hydrostatic pressure and the formation pore pressure results in a pressure differential that drives the drilling fluid to invade the porous rock formation and form a filter cake on the wellbore. There are a large number of natural fractures in shale reservoirs, and the degree of fracture development directly determines the reservoir quality and the level of shale oil and gas production. There are two main types of natural shale fractures: tectonic and non-tectonic fractures [74]. These fractures make it very easy for drilling fluid to penetrate into the formation under the action of pressure difference and cause a reaction with clay minerals. Compared to sandstones or limestones, shale formations are characterized by an abundance of clay minerals. Therefore, preventing water in the drilling fluid from penetrating into the deep reservoir through fractures is an important task to ensure the wellbore integrity. The main function of mechanical inhibitors in the drilling process is physical plugging. The basic principle of physical plugging is to seal the pores and throats of formations with solid particles or gel to prevent drilling fluid from invading the shale. The theories related to the physical plugging mechanism of drilling fluid are mainly concerned with the plugging effect and the bridging process [75]. When the size of the solid particles in the drilling fluid is comparable to the pore size of the formation, the solid particles are inserted into the pores and result in the plugging effect. The bridging process refers to the aggregation of multiple solid particles or gel particles to block the pores in the rock matrix that are larger than the size of a single solid particle. Figure 4 reveals the process of mechanical inhibitors plugging tiny pores, sealing microcracks, and forming an impermeable mud cake.

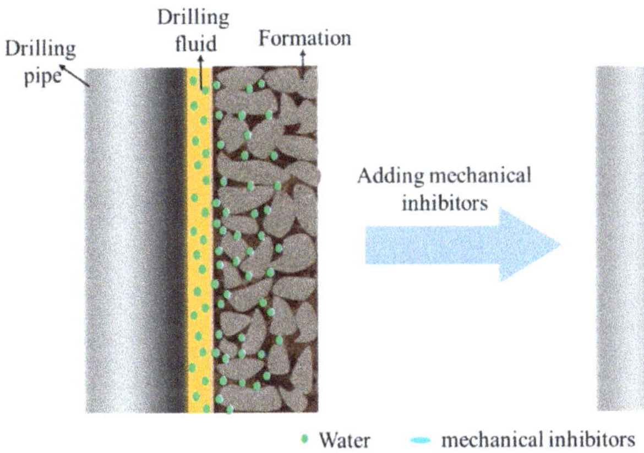

Figure 4. Inhibition mechanism of typical mechanical inhibitors.

Researchers have developed some inert materials and applied them to WBDFs to achieve physical plugging of formations [76–78]. Some novel materials, such as graphene nanoparticles, can not only have a plugging effect in the shale matrix, but can also improve the rheology, filtration, and lubricity of drilling fluids [79]. Asphaltene and gilsonite are two commonly used materials for plugging larger formation pore throats and microfractures [80,81]. However, these materials cannot be applied individually and often need to be combined with other suitable loss circulation materials of different sizes and with different properties in order to realize the bridging process and ultimately seal the cracks. Yang et al. [82] observed the surface of shale blocked by nanoparticles using SEM technology and proposed that shale has a wide range of throat sizes, and that nanoparticles preferentially blocked the throats closest to their own size. Therefore, the most critical principle for physical plugging of formation fractures and pores is to select solid particles with appropriate particle size.

5. Mechanical Inhibitors

Shale is an inherently heterogeneous system consisting of organic and inorganic matrices [83]. Microporous (<2 nm) and mesoporous (2~50 nm) structures are the dominant pore size distributions in the organic matrix. The inorganic matrix mainly comprises some clay minerals; the mesopores (2~50 nm) and macropores (>50 nm) are the main matrix pore structures [84]. Mechanical inhibitors do not react chemically with the swelling clay minerals in shale, and mainly block shale pores and cracks by physical means. Therefore, the key to physical plugging with mechanical inhibitors is for the size of inhibitor particles to match the size distribution of shale pores and fractures. Silicon-based, carbon-based, metal-based, and mineral-based nanoparticles have been tested as shale mechanical inhibitors in WBDFs [85]. The commonly used mechanical shale inhibitors are shown in Figure 5.

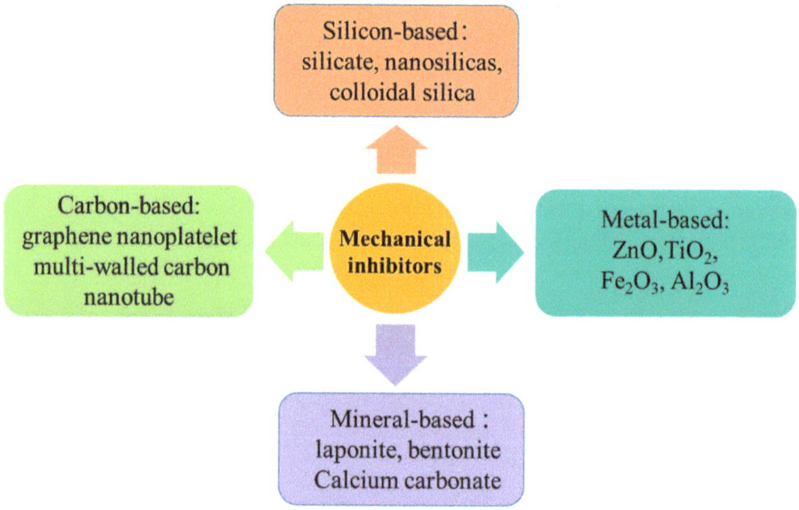

Figure 5. Classification and typical representatives of shale mechanical inhibitors.

5.1. Silicon-Based

Nanosilica is the most frequently employed mechanically inhibited silica-based particle in WBDFs. The potential of nanosilica as an inhibitor lies primarily in the physical plugging of tiny pores in shale. Silicon-based inhibitors can physically occupy the pore spaces within shale formations. By filling these pores, they reduce the permeability of the rock, limiting the flow of fluids that could lead to instability or swelling. However, when nanosilica is thoughtfully functionalized, the inhibition mechanism will go beyond plugging and may involve hydrogen bonding, and the inhibition potential of the modified nanosilica will be further strengthened. Xia et al. [86] investigated the effect of nanosilica on the performance of brine-based drilling fluids and showed that a 3% concentration of nanosilica significantly increased the viscosity of drilling fluids, reduced filtration, and improved shale stability under specific salt conditions. When nanosilica is added in WBDFs, it can form a three-dimensional network within the fluid. This network restricts the movement of fluid molecules and increases the viscosity of the drilling fluid. The nanoparticles can also act as bridging agents, linking larger particles and can impart viscoelastic properties to WBDFs [87]. The application of mesoporous silica in WBDFs can significantly improve the shale recovery rate and reduce the filtration loss of drilling fluid under HTHP conditions [88]. Nanosilica stabilized micro-foam drilling fluid for coalbed methane extraction has been reported [89,90]. The synergistic effect between the sealing effect of foam in the coal matrix and the plugging effect of nanosilica on nanopores can minimize the invasion of drilling fluid filtrate into the coal rock, thereby reducing formation damage. The test results from Bardhan et al. [88] show that under high temperature and high pressure, adding nanosilica to WBDFs can reduce the filtration loss by 48% and increase the shale recovery rate by 60%. Kang et al. [91] used SEM to visualize the plugging behavior of silica. SEM analysis of the shale surface exhibited that the shale has a wide range of pore throats, and the nanoparticles mainly plugged those pore throats that met their size (Figure 6a). However, some nanoparticles can aggregate together in some cases to block larger pore throats (Figure 6b).

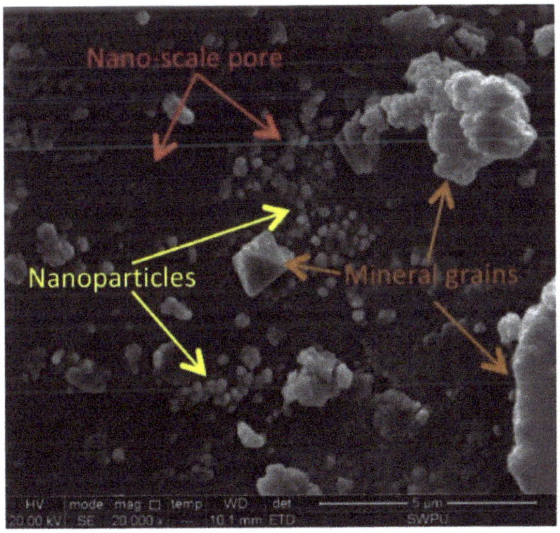

(a)

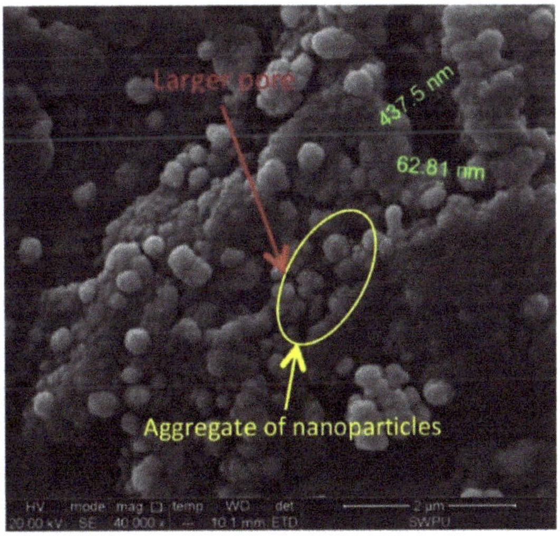

(b)

Figure 6. SEM images of (**a**) nanoparticles in shale surface and (**b**) nanoparticles aggregate plugging a pore throat [91].

The ability of silica to inhibit shale hydration and expansion by physical plugging is limited. Surface modification of silica to exert synergistic effects is an effective way to improve its inhibition ability. The recent advancement is that hydrophobically modified nanosilica has been reported to stop the hydration of clay minerals and ensure wellbore stability. The hydrophobic nanosilica prepared by the sol-gel method using nanosilica and silane coupling agent (dodecyltriethoxysilane, trimethylmethoxysilane, triethoxyoctylsilane, 3-aminopropyltriethoxysilane, and tetraethoxysilane) as raw materials exhibits excellent inhibition performance [92–94]. Additionally, nanosilica/polymer composites, which were obtained by free radical polymerization, demonstrated promising inhibitory properties and were resistant to high temperatures, up to 150 °C [95]. Some more complex composite hybrid materials, such as temperature-sensitive polymer-based nano-SiO_2 composites (SNAS), are employed to improve shale stability in WBDFs. The temperature sensitivity of SNAS allows them to transition from hydrophilic to hydrophobic at elevated temperatures, which enhances their ability to plug nano- and microfractures in the shale. This makes them highly effective under challenging drilling conditions [96]. The inhibition mechanism of nanosilica and its derivatives mainly involves three aspects, namely, the physical plugging of tiny cracks and pores by nanoparticles, the electrostatic interaction between the functional groups on the modified products and clay minerals, as well as the change in the shale surface wettability caused by the derivatives. Physical plugging by nanosilica generally needs to be combined with other inhibition mechanisms to achieve the maximum inhibitory potential. A detailed summary of the reported applications of nanosilica mechanical inhibitors is shown in Table 2.

Table 2. Summary of results for the performance of silicon-based mechanical inhibitors in WBDFs.

Composition	WBDFs Used	Dosage (wt. %)	Advantage	Limitation	Mechanism	Ref.
Nanosilica	Bentonite; Xanthan gum; modified starch; NaOH	1.0~5.0	Reduce filtration; improve the performance of saline drilling fluids	High concentrations have a negative impact on inhibition	Physical plugging	[86]
Mesoporous nanosilica	Bentonite Xanthan gum; polyanionic cellulose; KCl; KOH; N-octyl alcohol	0.05~1.0	Under HTHP, 1% mesoporous nanosilica can reduce filtration loss by 48% and increase shale recovery by 60%.	The HTHP filtration loss is large after hot rolling	Physical plugging	[88]
Colloidal silica	Typical field drilling mud filtrates	5.0~30	68% reduction in fluid penetration compared to field drilling fluids	High concentration requirement (20%)	Physical plugging	[97]
Superhydrophobic nanosilica (Modified by silane coupling agent)	Na-bentonite	1.0~3.0	Compared with KCl (3%), the linear expansion rate is reduced by 67%, and the wetting is reversed.	The effect of temperature on inhibition was not addressed	Physical plugging; electrostatic interaction and wettability alteration	[92]
Superhydrophobic nanosilica (modified by perfluorooctane triethoxysilane)	Bentonite	0.5~3.0	Better inhibition than KCl and polyamine; improve shale strength at high concentrations	-	Physical plugging; reduced surface energy; wettability alteration	[93]
Hydrophobic nanosilica (modified by stearyl trimethyl ammonium chloride)	Bentonite	0.5~3.0	Compared with KCL, the shale cuttings recovery rate is increased by 74% and the shale surface free energy is reduced by 80%	Poor linear swelling results	Physical plugging; reduced surface energy; wettability alteration	[98]
Nanosilica/ cationic polymer composite	Bentonite	0.1~2.0	Better than KCl and polyamine; better dispersion	-	Physical plugging; adsorption; compression double electric layer	[95]

5.2. *Carbon-Based*

Carbon-based nanoparticles, such as graphene nanosheets and carbon nanotubes, are effective in inhibiting shale expansion. These materials can physically occupy and block the pores in shale, thereby reducing fluid invasion and minimizing the hydration of clay minerals. By integrating into the shale structure, carbon-based inhibitors enhance the overall mechanical strength of the shale [99]. Their incorporation also decreases the permeability of shale, which limits the penetration of water and other drilling fluids that may cause swelling. Some new carbon-based nanoparticles can be obtained by chemically modifying graphene nanosheets, such as graphene oxide (GO) and carbonitride. As a new type of nanomaterial, GO consists of densely-packed carbon atoms forming a honeycomb lattice structure, which has a significant impact on the rheology, lubricity, and filtration control of drilling fluids [100]. The introduction of GO contributes to the formation of a more compact and thinner filter cake to reduce fluid loss [101]. Rana et al. [102] reported the application of GO-grafted glucopyranose in inhibiting shale expansion, and the modified GO prevents water molecules from penetrating into clay minerals through plugging and hydrogen bonding. Lv et al. [103] synthesized Janus amphiphilic graphene oxide (JAGO) using the Pickering emulsion method and modified one side of the GO sheet with dodecylamine. The two-dimensional nanosheet amphiphilic structure of JAGO enables it to have both chemical inhibition and physical plugging effects. The hydrophilic side of the JAGO could spontaneously adhere to the clay surface, and the outward hydrophobic surface forms a hydrophobic shield that prevents clay–water interactions (Figure 7).

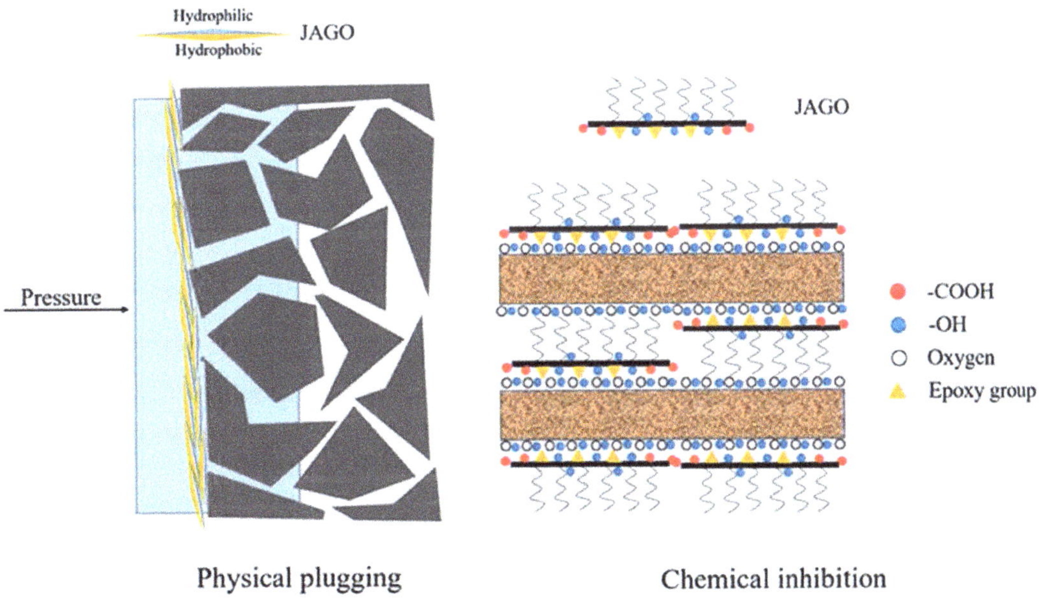

Figure 7. The physical and chemical inhibition mechanisms of JAGO [103].

Another type of carbon-based nanoparticle, carbon nanotubes (CNTs), can be used as a plugging agent in WBDFs to control drilling fluid loss due to its excellent mechanical and thermal properties. CNTs are a substance with the same chemical composition as carbon atoms and have a unique cylindrical physical structure [104]. Okoro et al. [105] found that carbon nanotubes can control filtration of both WBDFs and OBDFs, but the optimal concentrations for use are different. Liu et al. [106] investigated the effect of CNTs on the filtration of drilling fluid under high-temperature and high-salt conditions, and conducted a microscopic analysis of the drilling fluid mud cake. The results indicated that 6.1 μm pores were present in the mud cake prior to the addition of CNTs (Figure 8a). However, following the incorporation of CNTs, the pore size in the mud cake significantly decreased, resulting in a smoother and denser structure (Figure 8b). This improvement can be attributed to the adsorption of CNTs onto the surfaces of bentonite particles, which enhances spatial resistance between the particles, thereby promoting effective dispersion even under high-temperature and high-salinity conditions.

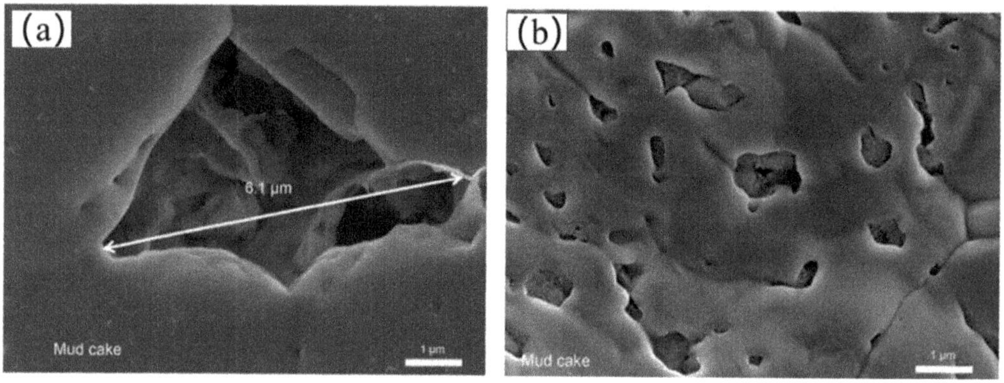

Figure 8. (**a**) SEM images of cakes of 4% bentonite + 10% NaCl after hot rolling for 16 h at 180 °C and (**b**) SEM images of cakes of 4% bentonite + 10% NaCl + 0.05% CNTs after hot rolling for 16 h at 180 °C [106].

Previous studies have attempted to improve the colloidal stability and dispersion of CNTs in electrolytes through the addition of surfactants and polymer, such as polyvinylpyrrolidone, cetyltrimethylammonium bromide, and polyethylene glycol [107,108]. However, the good colloidal stability of CNTs is not easy to achieve because it has weak interactions with surfactants. Table 3 demonstrates the application of carbon-based mechanical inhibitors in WBDFs in recent years.

Table 3. Application of carbon-based mechanical inhibitors in WBDFs.

Composition	WBDFs Used	Dosage (wt. %)	Advantage	Limitation	Mechanism	Ref.
graphene oxide (GO)	Xanthan gum; carboxymethyl cellulose	0.05~0.15	Forming compact and thinner filter cake	The effect of temperature on the properties of GO in drilling fluids was not addressed	Physical plugging	[101]
GO grafted with glucopyranose	Bentonite; resinex; Xanthan gum; poly aluminum chloride;	0.85	Better than KCl and traditional drilling mud; 90% shale dispersion recovery	The optimal concentration of modified graphene has not been determined	Physical plugging and hydrogen bonding	[102]
Janus amphiphilic graphene oxide (JAGO)	Na-Bentonite; Na_2CO_3	0.01~0.2	Better than GO, KCl, PA, and nanosilica	-	Physical plugging; adsorption and hydrogen bonding	[103]
Amine-functionalized graphene oxide	Xanthan gum; NaOH; barite	0.5	the thinnest filter cake with the least filtrate volume	-	Physical plugging; hydrophobicity	[109]
Multi-walled carbon nanotubes (MWCNT)	NaCl; $CaCl_2$	0.1~0.3	Better salt tolerance and thermal stability; excellent plugging performance	Thicker filter cake	Physical plugging; bridge	[110]

5.3. Metal-Based

In recent years, metal-based nanoparticles have been popularly used to improve filtration properties of drilling fluids due to their nanopore plugging effect. These particles mainly involve ZnO, CuO, TiO_2, Fe_3O_4, and MnO_2. These metal salts dissociate in solution, releasing metal ions that can interact with clay particles and shale surfaces. The presence of metal ions can change the surface charge of clay minerals in shale, leading to enhanced flocculation and reducing the amount of free water in the pore spaces [111]. Metal nanoparticles can penetrate the micro- and nano-scale pores in shale formations, filling them and effectively reducing permeability [112]. Dejtaradon et al. [113] found that the addition of ZnO and CuO to WBDFs significantly improves its rheological and filtration properties, especially at elevated temperatures and pressures. CuO nanoparticles provide better fluid loss reduction than ZnO. Bayat et al. [114] reported the effects of TiO_2 and ZnO on drilling fluid loss and demonstrated the relationship of temperature on the plugging ability of metal-based nanoparticles. Some metal-based nanoparticles can reduce the tendency of drilling fluid properties to thermally degrade, for example, MnO_2, which improves the rheological and filtration properties of the mud system by imparting thermal stability along with minimizing filtration losses [115]. Ejtemaee et al. [116] evaluated the effects of Al_2O_3, TiO_2, and Fe_3O_4 on the performance of drilling fluids, and indicated that Fe_3O_4 displayed appropriate rheology and filtration, and caused minimal damage to the formation.

A study on iron oxide demonstrated that it has great application potential beyond mechanical inhibitors in WBDFs. The remarkable magnetic characteristics of iron oxide particles can be used to monitor and control the properties of drilling fluids in real time, which makes it possible to develop smart drilling fluids using iron oxide particles [117]. Furthermore, filtration test results of iron oxide particles in WBDFs demonstrate that iron-based nanoparticles more effectively reduce fluid loss under HPHT conditions compared to low-pressure and low-temperature (LPLT) conditions [118,119]. In an attempt to plug the pores, iron oxide particles are forced into smaller pores under hydrostatic pressure, thereby reducing filtration loss and improving the inhibition (Figure 9). Compared with other metal-based nanoparticles, ZnO has advantages in terms of environmental friendliness and multifunctionality. ZnO can significantly enhance the viscosity and gel strength of WBDFs, and can effectively remove H_2S, which helps to improve the safety of drilling operations [120,121]. Additionally, the acute toxicity test of ZnO revealed that the survival rate of white leg shrimp in WBDFs waste was 100%, which proved its environmental friendliness [122]. Medhi et al. [123]'s research shows that drilling fluid containing zinc oxide nanoparticles can reduce fluid loss by about 50% compared with traditional formulas. Other metal-based nanoparticles, such as copper oxide and titanium dioxide, also exhibited good inhibitory abilities, reducing the water absorption of shale samples by 83% and 54%, respectively [124].

5.4. Mineral-Based

Bentonite, a widely recognized mineral primarily composed of MMT, has been a key component in WBDFs since the 19th century. It serves a dual function, both regulating rheological properties and minimizing filtration loss. Bentonite's excellent swelling ability allows it to form a low-permeability mud cake in the wellbore. When hydrated, it absorbs water and expands to several times its dry volume, effectively filling pores and fractures in the formation to prevent fluid loss and stabilize the wellbore [125–127]. Its colloidal properties produce a gel-like structure that enhances cuttings suspension and reduces drilling fluid intrusion into porous formations [128,129]. In addition, the fine particles of bentonite can fill the pore space of shale formations, reduce mud cake permeability, and limit fluid flow. Bentonite can also undergo cation exchange, thereby changing the electrostatic interaction on the shale surface and further enhancing mechanical stability [130]. These properties render bentonite particularly effective in controlling lost circulation and maintaining wellbore stability. Another material frequently employed in WBDFs for plugging and bridging is calcium carbonate, whose particle size distribution is critical to the effectiveness of mechanical inhibition of drilling fluids [131,132]. For fractured reservoirs, drilling fluid plugging technology mainly utilizes bridging particles such as calcium carbonate in drilling fluids to transform fractures into pores and form a protective barrier, thereby preventing fluid loss and maintaining wellbore stability. Additionally, calcium carbonate has acid-soluble properties, which can improve the acid washing effect of drilling fluid from the perspective of reservoir protection [133–136]. The application of graded calcium carbonate bridging particles in WBDFs has demonstrated that combining particles of varying types and sizes can enhance the plugging and bridging, improving drilling fluids performance across diverse drilling conditions [137,138].

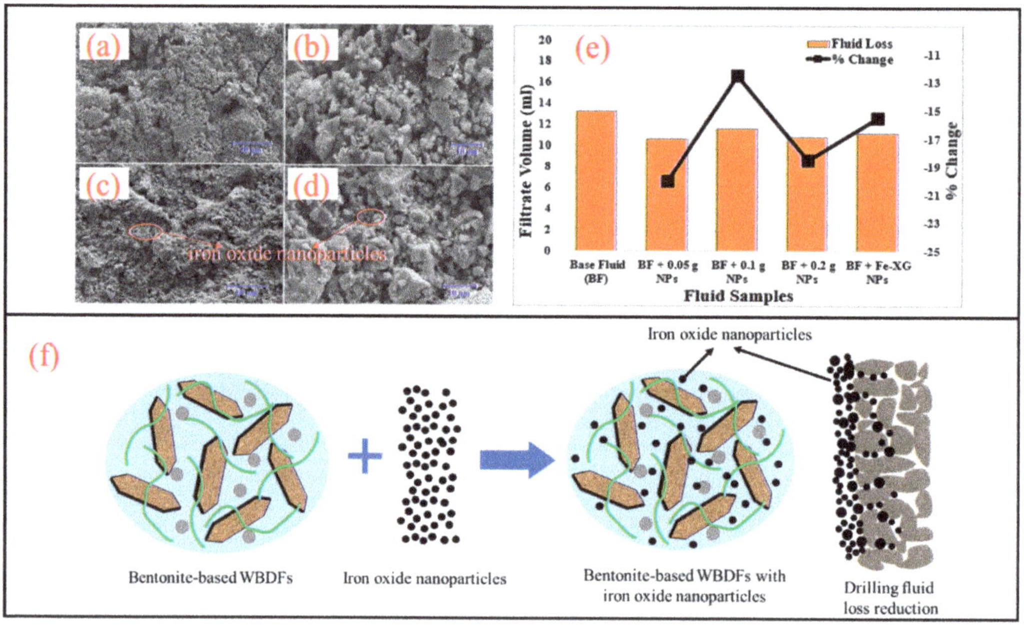

Figure 9. SEM images of filter cake for Bentonite-based fluids (**a**,**b**); SEM images of filter cake for Bentonite-based fluids with iron oxide nanoparticles (**c**,**d**); filtrate volume after fluid loss test of Bentonite-based fluids containing iron oxide nanoparticles (**e**); and schematic diagram of iron oxide particles being pushed into the filter cake and plugging the pores (**f**) (Modified from [118]).

In addition to natural mineral-based materials, there has been increasing interest in the application of synthetic mineral materials as mechanical inhibitors in WBDFs [139,140]. Laponite is a synthetic material with a structure and composition similar to natural hectorite, and which belongs to the family of phyllosilicates built up of silica tetrahedron and magnesia octahedron at a ratio of 2:1 along the Z axis. Its crystals have a disk-like structure with diameters of 25 nm and thicknesses of 0.92 nm [141,142]. Laponite inhibits shale hydration by plugging interlayer spaces in clay through electrostatic interactions. Its excellent thixotropic properties in WBDFs allow it to form a protective film on the wellbore, further reducing water penetration. Additionally, the ultra-fine laponite particles are small enough to seal the nano-scale pores in shale, preventing water invasion and stabilizing the wellbore effectively. The results of SEM showed that the original shale fragments were porous and loose, with many pores and cracks. Water from drilling fluid can easily penetrate into the shale matrix, causing wellbore instability. After being treated with the laponite suspension, the surface of the shale sample became seamless, and the majority of pores and fractures disappeared, which helped prevent water penetration (Figure 10) [143]. These combined mechanisms make laponite a more effective shale inhibitor compared to traditional additives like KCl and polyester amine [143,144].

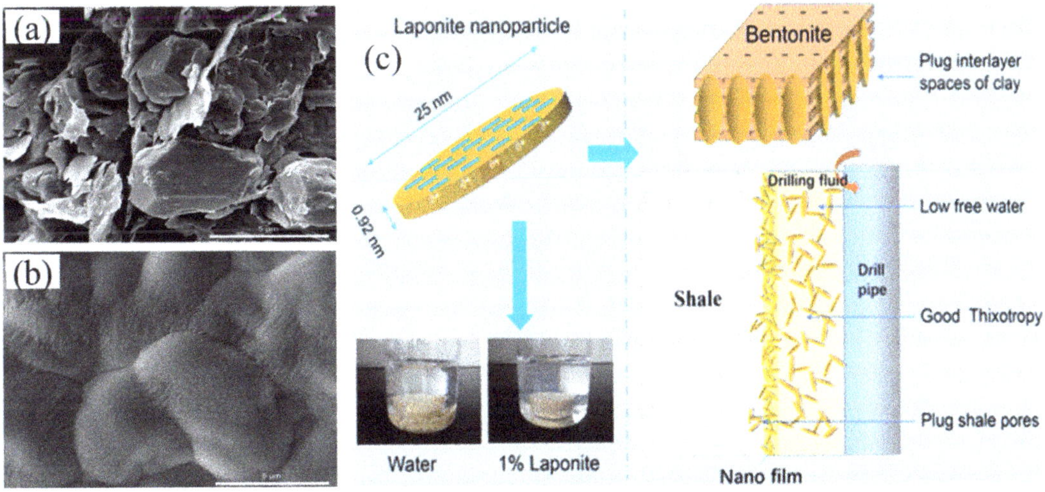

Figure 10. SEM images of original shale fragment (**a**); recovered shale fragments treated with 2 wt% laponite in shale recovery test (**b**) [143]; and schematic diagram of the plugging and inhibition mechanism of laponite (**c**) [144].

Silicon-based, carbon-based, metal-based, and mineral-based materials are widely used as shale mechanical inhibitors and primarily play a role in plugging shale fractures and in preventing water penetration into the shale. In addition, other mechanical inhibitors have also been reported to seal shale pores. For example, carbon ash, a by-product of the steel recycling process, shows great potential in drilling fluids. Carbon ash adheres to bentonite surfaces in drilling fluids by electrostatic action, improving its network structure and ultimately enhancing the filtration properties of drilling fluids [145,146]. Perlite is also used as an additive to control the fluid loss potential of high-density drilling fluids. The perlite particles form a sedimentary layer that plugs the pores in the filter cake, reducing the thickness of the filter cake formed by the drilling fluid by 45% [147,148]. Mixed metal hydroxides (MMHs) are often used as inorganic viscosifiers in drilling fluids due to their unique electrostatic interactions. Electrostatic attraction occurs between these particles and the positively charged bentonite surface and significantly increases the filtrate viscosity to prevent water from entering the formation [149]. Table 4 summarizes the application characteristics of common shale mechanical inhibitors.

Table 4. Comparison of shale mechanical inhibitors in terms of mechanisms, cost, base fluid, advantages, limitations, and application areas.

Shale Mechanical Inhibitors	Mechanisms	Cost	Base Fluid	Advantages	Limitations	Application Areas	Ref.
Silicon-based	Physical plugging; reduced surface energy; wettability alteration	Moderate	Water/oil	High thermal stability; low reactivity; viscosity control	Rigid particles; poor deformability	High temperatures and pressures; fracture shale reservoir	[88,91,150]
Carbon-based	Physical plugging; adsorption and hydrogen bonding	High	Water/oil	Low density; environmental compatibility; enhanced lubrication	Lower thermal stability; high cost	Moderate temperatures; porous shale reservoir; gypsum-salt rock	[63,106,110]
Metal-based	Physical plugging; ionic interaction	High	Water/oil	Good solubility; plugging nano-micro pores	High cost; the effectiveness varies significantly	Moderate to high temperatures; porous shale reservoir	[121,123]

Table 4. *Cont.*

Shale Mechanical Inhibitors	Mechanisms	Cost	Base Fluid	Advantages	Limitations	Application Areas	Ref.
Mineral-based	Physical plugging; electrostatic interactions	Low	Water/oil	Cost-effective; compatibility	Limited performance; weak temperature resistance	Moderate to high temperatures; fracture shale reservoir	[127,151]

6. Challenges

Shale mechanical inhibitors offer several advantages in drilling operations, particularly in enhancing wellbore stability in challenging shale formations. By reinforcing the mechanical integrity of the shale, these inhibitors help mitigate issues such as wellbore collapse and swelling, ultimately leading to reduced operational costs [152]. Shale mechanical inhibitors provide significant benefits for wellbore stability but also face notable challenges in their application. Compatibility is a primary concern, as the introduction of these inhibitors can interact negatively with other additives in drilling fluid systems, potentially affecting the overall performance and properties [153]. Additionally, cost and environmental protection are also vital considerations. Finally, the development of intelligent drilling fluids poses additional complexities. The application challenges of shale mechanical inhibitors are summarized in Figure 11.

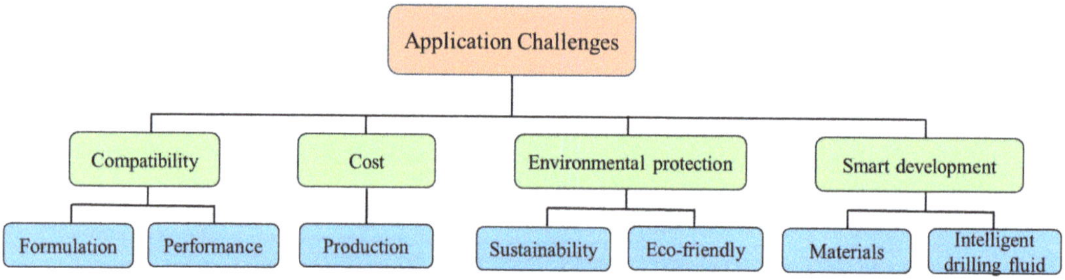

Figure 11. The application challenges of shale mechanical inhibitors.

6.1. Compatibility of Materials and Drilling Fluid Performance

Shale mechanical inhibitors, including silicon-based and carbon-based materials, have demonstrated promising results in reducing shale swelling and enhancing wellbore stability [154]. However, their effectiveness can be compromised by interactions with other fluid components, such as precipitation resulting from physical or chemical interactions. Furthermore, the variability in the chemical structure of silicates may influence their performance in shale formations, necessitating customized formulations tailored to specific geological conditions. In the case of carbon-based mechanical inhibitors, their performance may diminish or even degrade under high-temperature conditions, leading to a significant decline in the overall performance of the drilling fluid system [155]. Therefore, innovative solutions are essential for effectively integrating these materials into the fluid system.

6.2. Cost-Effectiveness

The economic feasibility of incorporating advanced mechanical shale inhibitors into WBDFs is an essential consideration for operators and companies. Although these inhibitors have the potential to significantly improve wellbore stability and overall drilling efficiency, their high initial cost may hinder widespread application. Effective inhibitors can decrease the incidence of wellbore instability, resulting in fewer drilling interruptions, reduced NPT, and the lower operational costs associated with remedial measures [156]. However, the synthesis, processing, and specialized treatment of these advanced inhibitors entail higher upfront research and development expenses. It is, therefore, essential to thoroughly evaluate the relationship between the performance indicators of drilling fluids and the return on the investment cost of inhibitors. By comparing these metrics to historical performance data for conventional inhibitors, the potential return on investment associated with advanced inhibitor materials can be better understood.

6.3. Environmental Regulations

The increasing emphasis on environmental sustainability and regulatory compliance complicates the development and application of mechanical shale inhibitors. Environmental regulation aims to mitigate the impact of drilling activities on the environment, addressing concerns such as groundwater contamination, emissions, and drilling waste management [77]. In this context, selecting appropriate mechanical shale inhibitors is crucial, as traditional materials may pose environmental risks, including toxicity to aquatic life and long-term ecological damage. The development of eco-friendly inhibitors, such as those derived from natural minerals, biodegradable materials, or low-toxicity chemical formulations, has emerged as a key focus in both research and industry practice [39]. These materials not only help meet compliance standards, but also improve the overall sustainability of drilling operations, which is increasingly important in a world striving for greener energy solutions.

6.4. Smart Development of Mechanical Inhibitors

The integration of smart materials into the field of shale mechanical inhibitors represents a transformative approach that promises to significantly improve the performance of WBDFs. Smart materials are carefully engineered to dynamically respond to environmental changes such as pressure, temperature, magnetism, or chemical composition, thereby adjusting their properties in real time. An example with significant application prospect is the use of iron oxide-based magnetic materials as mechanical inhibitors to formulate smart drilling fluids [157]. These materials offer excellent advantages over traditional inhibitors in drilling operations, particularly in addressing the challenges posed by unstable shale formations. For example, the use of encapsulated materials to release active ingredients in response to external stimuli, such as temperature or pressure changes, can improve the reliability and effectiveness of shale inhibition strategies [96]. Currently, the latest research area of drilling fluids aims to create fluids that can autonomously identify complex geological conditions and automatically adjust their properties to mitigate or prevent drilling accidents. To be precise, the design and implementation of intelligent drilling fluid technologies characterized by "self identification, self-tuning, and self-adaptation" will fundamentally address the challenges in drilling fluid technology [117,158]. Additives play a critical role in enabling drilling fluids to self-adjust their performance. The smart shale mechanical inhibitors, such as some temperature-responsive, pressure-responsive, and shape-memory materials [159,160], can quickly detect changes in their external environment and dynamically modify their properties to achieve more efficient inhibition of shale.

7. Conclusions and Perspectives

In this paper, clay minerals associated with wellbore instability are presented and the types, mechanisms, and effectiveness of shale mechanical inhibitors are summarized. The following conclusions are obtained:

(I) The hydration and expansion of clay minerals are key contributors to wellbore instability during drilling operations. Effective inhibitors can stabilize the wellbore by minimizing the interaction between drilling fluids and clay, thereby preserving wellbore integrity and ensuring safer, more efficient drilling in challenging formations.

(II) Mechanical inhibitors have demonstrated considerable potential in reinforcing wellbore stability, particularly in shale formations where chemical inhibitors alone are often insufficient. These inhibitors operate through physical mechanisms, primarily by plugging cracks and sealing microcracks, thereby reducing fluid intrusion and enhancing structural integrity.

(III) Despite the benefits of mechanical inhibitors, their widespread application encounters challenges in terms of compatibility with other drilling fluid components and cost-effectiveness. Interactions between mechanical inhibitors and drilling fluid additives may diminish the inhibitor's efficacy, highlighting the need for customized drilling fluid formulations.

(IV) The future of mechanical inhibitors is rooted in the innovation of smart and environmentally sustainable materials. Research should prioritize the synthesis of smart inhibitors that can dynamically respond to environmental changes such as pressure, temperature, and magnetism to enhance their adaptability in stabilizing the wellbore.

Author Contributions: Conceptualization, E.A. and Z.L.; methodology, E.A.; formal analysis, B.Z. and L.W.; investigation, E.A. and Z.L.; data curation, C.M.; writing—original draft preparation, E.A.; writing—review and editing, E.A. and Q.L.; visualization, E.A. and Q.L.; supervision, Q.L.; funding acquisition, Q.L. and Z.L. All authors have read and agreed to the published version of the manuscript.

Funding: This work was funded by National Natural Science Foundation of China (52374009), China Postdoctoral Science Foundation (2023MD734222), Postdoctoral Research Project of Shaanxi Province (2023BSHEDZZ319).

Conflicts of Interest: Bo Zhang is employed by Department of Mahu Area of Xinjiang Oilfield Company. The paper reflects the views of the scientists and not the companies.

References

1. Ma, T.; Zhang, Y.; Qiu, Y.; Liu, Y.; Li, Z. Effect of parameter correlation on risk analysis of wellbore instability in deep igneous formations. *J. Pet. Sci. Eng.* **2022**, *208*, 109521. [CrossRef]
2. Li, Q.; de Viguerie, L.; Laporte, L.; Berraud-Pache, R.; Zhuang, G.; Souprayen, C.; Jaber, M. Oily bioorganoclays in drilling fluids: Micro and macroscopic properties. *Appl. Clay Sci.* **2024**, *247*, 107186. [CrossRef]
3. Liu, W.; Lin, H.; Liu, H.; Luo, C.; Wang, G.; Deng, J. Numerical Investigation of Wellbore Stability in Deepwater Shallow Sediments. *Geofluids* **2021**, *2021*, 5582605. [CrossRef]
4. Pandya, S.; Ahmed, R.; Shah, S. Wellbore Cleanout in Inclined and Horizontal Wellbores: The Effects of Flow Rate, Fluid Rheology, and Solids Density. *SPE Drill. Complet.* **2020**, *35*, 48–68. [CrossRef]
5. Aslannezhad, M.; Keshavarz, A.; Kalantariasl, A. Evaluation of mechanical, chemical, and thermal effects on wellbore stability using different rock failure criteria. *J. Nat. Gas Sci. Eng.* **2020**, *78*, 103276. [CrossRef]
6. Liu, C.; Han, Y.; Phan, D.T.; Abousleiman, Y.N. Stress solutions for short- and long-term wellbore stability analysis. *J. Nat. Gas Sci. Eng.* **2022**, *105*, 104693. [CrossRef]
7. Ibrahim, A. A review of mathematical modelling approaches to tackling wellbore instability in shale formations. *J. Nat. Gas Sci. Eng.* **2021**, *89*, 103870. [CrossRef]
8. Kang, Y.; Yu, M.; Miska, S.; Takach, N.E. Wellbore Stability: A Critical Review and Introduction to DEM. In Proceedings of the SPE Annual Technical Conference and Exhibition, New Orleans, LA, USA, 4–7 October 2009; p. SPE–124669-MS.
9. Halafawi, M.; Avram, L. Borehole insitu stress stability analysis of RBS-9 field utilizing the inversion technique. *Tech. Sci.* **2019**, *4*, 67–78. [CrossRef]
10. Wang, M.; Li, M.; Li, J.-B.; Xu, L.; Zhang, J.-X. The key parameter of shale oil resource evaluation: Oil content. *Pet. Sci.* **2022**, *19*, 1443–1459. [CrossRef]
11. Babatunde, K.A.; Negash, B.M.; Jufar, S.R.; Ahmed, T.Y.; Mojid, M.R. Adsorption of gases on heterogeneous shale surfaces: A review. *J. Pet. Sci. Eng.* **2022**, *208*, 109466. [CrossRef]
12. Du, J.; Whittle, A.J.; Hu, L.; Divoux, T.; Meegoda, J.N. Characterization of meso-scale mechanical properties of Longmaxi shale using grid microindentation experiments. *J. Rock Mech. Geotech. Eng.* **2021**, *13*, 555–567. [CrossRef]
13. Liu, H.; Feng, X.; Liu, L.; Li, T.; Tang, C. Mechanical properties and failure characteristics of anisotropic shale with circular hole under combined dynamic and static loading. *Int. J. Rock Mech. Min. Sci.* **2023**, *170*, 105524. [CrossRef]
14. Basfar, S.; Bageri, B.; Elkatatny, S. Effect of Qusaiba shale formation on high-pressure high-temperature drilling fluids properties. *Geoenergy Sci. Eng.* **2023**, *224*, 211608. [CrossRef]
15. Wang, B.; Sun, J.; Shen, F.; Li, W.; Zhang, W. Mechanism of wellbore instability in continental shale gas horizontal sections and its water-based drilling fluid countermeasures. *Nat. Gas Ind. B* **2020**, *7*, 680–688. [CrossRef]
16. Paswan, B.K.; Mahto, V. Development of environment-friendly oil-in-water emulsion based drilling fluid for shale gas formation using sunflower oil. *J. Pet. Sci. Eng.* **2020**, *191*, 107129. [CrossRef]

17. Li, Q.; de Viguerie, L.; Souprayen, C.; Casale, S.; Jaber, M. Oil-based drilling fluid inspired by paints recipes. *Appl. Clay Sci.* **2023**, *245*, 107120. [CrossRef]
18. Alves, G.M.; Petri Júnior, I. Microwave remediation of oil-contaminated drill cuttings—A review. *J. Pet. Sci. Eng.* **2021**, *207*, 109137. [CrossRef]
19. Luo, Z.; Wang, L.; Yu, P.; Chen, Z. Experimental study on the application of an ionic liquid as a shale inhibitor and inhibitive mechanism. *Appl. Clay Sci.* **2017**, *150*, 267–274. [CrossRef]
20. Saleh, T.A. Advanced trends of shale inhibitors for enhanced properties of water-based drilling fluid. *Upstream Oil Gas Technol.* **2022**, *8*, 100069. [CrossRef]
21. Jia, H.; Huang, P.; Han, Y.; Wang, Q.; Jia, K.; Sun, T.; Zhang, F.; Yan, H.; Lv, K. Investigation for the novel use of a typical deep eutectic solvent as a potential shale inhibitor. *Energy Sources Part A Recovery Util. Environ. Eff.* **2022**, *44*, 1402–1415. [CrossRef]
22. Ghasemi, A.; Jalalifar, H.; Norouzi Apourvari, S.; Sakebi, M.R. Mechanistic study of improvement of wellbore stability in shale formations using a natural inhibitor. *J. Pet. Sci. Eng.* **2019**, *181*, 106222. [CrossRef]
23. Muhammed, N.S.; Olayiwola, T.; Elkatatny, S. A review on clay chemistry, characterization and shale inhibitors for water-based drilling fluids. *J. Pet. Sci. Eng.* **2021**, *206*, 109043. [CrossRef]
24. Ahmed, H.M.; Kamal, M.S.; Al-Harthi, M. Polymeric and low molecular weight shale inhibitors: A review. *Fuel* **2019**, *251*, 187–217. [CrossRef]
25. Jingyuan, M.; Boru, X.; Yuxiu, A. Advanced developments in low-toxic and environmentally friendly shale inhibitor: A review. *J. Pet. Sci. Eng.* **2022**, *208*, 109578. [CrossRef]
26. Muhammed, N.S.; Olayiwola, T.; Elkatatny, S.; Haq, B.; Patil, S. Insights into the application of surfactants and nanomaterials as shale inhibitors for water-based drilling fluid: A review. *J. Nat. Gas Sci. Eng.* **2021**, *92*, 103987. [CrossRef]
27. Hammad Rasool, M.; Ahmad, M.; Ayoub, M.; Zamir, A.; Adeem Abbas, M. A review of the usage of deep eutectic solvents as shale inhibitors in drilling mud. *J. Mol. Liq.* **2022**, *361*, 119673. [CrossRef]
28. Swai, R.E. A review of molecular dynamics simulations in the designing of effective shale inhibitors: Application for drilling with water-based drilling fluids. *J. Pet. Explor. Prod. Technol.* **2020**, *10*, 3515–3532. [CrossRef]
29. Al-Arfaj, M.K.; Amanullah, M.; Sultan, A.S.; Hossain, E.; Abdulraheem, A. Chemical and Mechanical Aspects of Wellbore Stability in Shale Formations: A Literature Review. In Proceedings of the Abu Dhabi International Petroleum Exhibition and Conference, Abu Dhabi, United Arab Emirates, 10–13 November 2014.
30. Aston, M.S.; Alberty, M.W.; McLean, M.R.; de Jong, H.J.; Armagost, K. Drilling Fluids for Wellbore Strengthening. In Proceedings of the IADC/SPE Drilling Conference, Dallas, TX, USA, 2–4 March 2004.
31. Li, B.; Feng, Y.; Ma, C.; Li, S.; Lai, C.; Zhang, S.; Su, F. Particle Size Optimization of Lost Circulation Materials: A Comprehensive Experimental Study. In Proceedings of the 57th U.S. Rock Mechanics/Geomechanics Symposium, Atlanta, GA, USA, 25–28 June 2023.
32. Luo, B.; Wong, G.K.; Han, Y. Modeling of dynamic bridging of solid particles in multiple propagating fractures. *Int. J. Solids Struct.* **2023**, *262–263*, 112078. [CrossRef]
33. Jaf, P.T.; Razzaq, A.A.; Ali, J.A. The state-of-the-art review on the lost circulation phenomenon, its mechanisms, and the application of nano and natural LCM in the water-based drilling fluid. *Arab. J. Geosci.* **2022**, *16*, 32. [CrossRef]
34. Lashkari, R.; Tabatabaei-Nezhad, S.A.; Husein, M.M. Evaluation of shape memory polyurethane as a drilling fluid lost circulation and fracture plugging material. *Geoenergy Sci. Eng.* **2023**, *222*, 211445. [CrossRef]
35. Zhang, B.; Gao, B.; Ma, W.; Wang, S.; Qi, W.; Wu, J.; Wang, R.; Zhou, Y.; Liu, Y. Different behavior of uranium(VI) on two clay minerals: Montmorillonite and kaolinite. *J. Radioanal. Nucl. Chem.* **2023**, *332*, 4029–4046. [CrossRef]
36. Larson, S.L.; Ballard, J.H.; Runge, K.A.; Zhang, H.; Breland, B.R.; Nick, Z.H.; Weiss, C.A.; Han, F.X. Adsorption and characterization of exopolysaccharides from Rhizobium tropici on clay minerals. *Carbohydr. Polym. Technol. Appl.* **2023**, *5*, 100314. [CrossRef]
37. Shi, K.-Y.; Chen, J.-Q.; Pang, X.-Q.; Jiang, F.-J.; Hui, S.-S.; Zhao, Z.-C.; Chen, D.; Cong, Q.; Wang, T.; Xiao, H.-Y.; et al. Wettability of different clay mineral surfaces in shale: Implications from molecular dynamics simulations. *Pet. Sci.* **2023**, *20*, 689–704. [CrossRef]
38. Kasprzhitskii, A.; Kruglikov, A.; Ermolov, Y.; Yavna, V.; Pleshko, M.; Lazorenko, G. Structure of bound water layer on montmorillonite surface: The role of trans- and cis-vacant sites. *Appl. Surf. Sci.* **2024**, *642*, 158565. [CrossRef]
39. Abdullah, A.D.; Ali, J.A.; Abdalqadir, M. Exploring the role of hydrophobic nanofluids in reducing shale swelling during drilling: A step towards eco-friendly and sustainable practices. *Colloids Surf. A Physicochem. Eng. Asp.* **2024**, *694*, 134164. [CrossRef]
40. Ahmad, H.M.; Murtaza, M.; Shakil Hussain, S.M.; Mahmoud, M.; Kamal, M.S. Performance evaluation of different cationic surfactants as anti-swelling agents for shale formations. *Geoenergy Sci. Eng.* **2023**, *230*, 212185. [CrossRef]
41. Worden, R.; Morad, S. *Clay Mineral Cements in Sandstones*; John Wiley & Sons: Hoboken, NJ, USA, 2009.
42. Jumaeva, D.; Toirov, O.; Okhunjanov, Z.; Raximov, U.; Akhrorova, R. Investigation of the adsorption of nonpolar adsorbate molecules on the illite surface. *J. Chem. Technol. Metall.* **2023**, *58*, 353–359. [CrossRef]
43. Rimkevich, V.; Pushkin, A.; Malovitskii, Y.N.; Eranskaya, T.Y.; Girenko, I. Physicochemical processes of the fluoride processing of aluminum ores. *Theor. Found. Chem. Eng.* **2010**, *44*, 844–852. [CrossRef]
44. Ennis-King, J.; LaForce, T.; Paterson, L.; Black, J.R.; Vu, H.P.; Haese, R.R.; Serno, S.; Gilfillan, S.; Johnson, G.; Freifeld, B.; et al. Stepping into the Same River Twice: Field Evidence for the Repeatability of a CO_2 Injection Test. *Energy Procedia* **2017**, *114*, 2760–2771. [CrossRef]

45. Ohazuruike, L.; Lee, K.J. A comprehensive review on clay swelling and illitization of smectite in natural subsurface formations and engineered barrier systems. *Nucl. Eng. Technol.* **2023**, *55*, 1495–1506. [CrossRef]
46. Bee, S.-L.; Abdullah, M.A.A.; Bee, S.-T.; Sin, L.T.; Rahmat, A.R. Polymer nanocomposites based on silylated-montmorillonite: A review. *Prog. Polym. Sci.* **2018**, *85*, 57–82. [CrossRef]
47. Pavlidou, S.; Papaspyrides, C.D. A review on polymer–layered silicate nanocomposites. *Prog. Polym. Sci.* **2008**, *33*, 1119–1198. [CrossRef]
48. Jankovič, L'.; Madejová, J.; Komadel, P.; Jochec-Mošková, D.; Chodák, I. Characterization of systematically selected organo-montmorillonites for polymer nanocomposites. *Appl. Clay Sci.* **2011**, *51*, 438–444. [CrossRef]
49. Faheem, U. Montmorillonite: An Introduction to Properties and Utilization. In *Current Topics in the Utilization of Clay in Industrial and Medical Applications*; Mansoor, Z., Ed.; IntechOpen: Rijeka, Croatia, 2018. [CrossRef]
50. Shadizadeh, S.R.; Moslemizadeh, A.; Dezaki, A.S. A novel nonionic surfactant for inhibiting shale hydration. *Appl. Clay Sci.* **2015**, *118*, 74–86. [CrossRef]
51. Zhang, F.; Sun, J.; Chang, X.; Xu, Z.; Zhang, X.; Huang, X.; Liu, J.; Lv, K. A Novel Environment-Friendly Natural Extract for Inhibiting Shale Hydration. *Energy Fuels* **2019**, *33*, 7118–7126. [CrossRef]
52. Li, X.; Jiang, G.; Yang, L.; Peng, S. Study of gelatin as biodegradable shale hydration inhibitor. *Colloids Surf. A Physicochem. Eng. Asp.* **2018**, *539*, 192–200. [CrossRef]
53. Fu, L.; Liao, K.; Ge, J.; He, Y.; Huang, W.; Du, E. Preparation and inhibition mechanism of bis-quaternary ammonium salt as shale inhibitor used in shale hydrocarbon production. *J. Mol. Liq.* **2020**, *309*, 113244. [CrossRef]
54. Hamza, A.; Hussein, I.A.; Mahmoud, M. Chapter 1—Introduction to reservoir fluids and rock properties. In *Developments in Petroleum Science*; Hussein, I.A., Mahmoud, M., Eds.; Elsevier: Amsterdam, The Netherlands, 2023; Volume 78, pp. 1–19.
55. Patel, A.; Stamatakis, E.; Young, S.; Friedheim, J. Advances in Inhibitive Water-Based Drilling Fluids–Can they Replace Oil-Based Muds? In Proceedings of the International Symposium on Oilfield Chemistry, Houston, TX, USA, 28 February–2 March 2007; p. SPE–106476-MS.
56. Khodja, M.; Amish, M.; Canselier, J.P.; Bergaya, F. Chapter 5—Borehole instability related to clay minerals. In *Clay Science in Drilling and Drilling Fluids*; Zhuang, G., Yuan, P., Eds.; Elsevier: Amsterdam, The Netherlands, 2024; Volume 11, pp. 109–160.
57. Khalifa, A.Z.; Cizer, Ö.; Pontikes, Y.; Heath, A.; Patureau, P.; Bernal, S.A.; Marsh, A.T.M. Advances in alkali-activation of clay minerals. *Cem. Concr. Res.* **2020**, *132*, 106050. [CrossRef]
58. Li, Y.; Liu, Q.; Liu, L.; Liu, L.; Hou, D.; Wu, Y. Effect of original crystal size of kaolinite on the formation of intercalation compounds of coal-measure kaolinite. *Mater. Today Commun.* **2023**, *35*, 106130. [CrossRef]
59. Neeraj, K.; Chandra, M. Basics of Clay Minerals and Their Characteristic Properties. In *Clay and Clay Minerals*; Gustavo Morari Do, N., Ed.; IntechOpen: Rijeka, Croatia, 2021. [CrossRef]
60. Kong, L.; Tang, J.; Luo, Y.; Yuan, F.; Lin, Y.; Tao, R. Construction and Evaluation of a Degradable Drilling Fluid for Underground Coalbed Methane Extraction Boreholes. *ACS Omega* **2024**, *9*, 10426–10439. [CrossRef]
61. Cliffe, S.; Young, S. Agglomeration and Accretion of Drill Cuttings in Water-Based Fluids. 2008. Available online: https://www.scribd.com/document/242013374/Agglomeration-and-Accretion-of-Drill-Cuttings-in-Water-Based-Fluids-pdf (accessed on 14 November 2024).
62. Alujas Diaz, A.; Almenares Reyes, R.S.; Hanein, T.; Irassar, E.F.; Juenger, M.; Kanavaris, F.; Maier, M.; Marsh, A.T.; Sui, T.; Thienel, K.-C.; et al. Properties and occurrence of clay resources for use as supplementary cementitious materials: A paper of RILEM TC 282-CCL. *Mater. Struct.* **2022**, *55*, 139. [CrossRef]
63. Boruah, J.S.; Chowdhury, D. Advances in Carbon Nanomaterial–Clay Nanocomposites for Diverse Applications. *Minerals* **2023**, *13*, 26. [CrossRef]
64. Heald, M.T.; Anderegg, R.C. Differential cementation in the Tuscarora sandstone [Virginia-West Virginia]. *J. Sediment. Res.* **1960**, *30*, 568–577. [CrossRef]
65. Pittman, E.D.; Lumsden, D.N. Relationship between chlorite coatings on quartz grains and porosity, Spiro Sand, Oklahoma. *J. Sediment. Res.* **1968**, *38*, 668–670. [CrossRef]
66. Zhou, X.; Liu, X.; Liang, L. Analysis of changes in shale mechanical properties and fault instability activation caused by drilling fluid invasion into formations. *J. Pet. Explor. Prod. Technol.* **2024**, *14*, 2343–2358. [CrossRef]
67. Yang, M.; Xu, Y.; Zhang, J.; Chen, H.; Liu, S.; Li, W.; Hao, Y. Near-Infrared Spectroscopic Study of Heavy-Metal-Contaminated Loess Soils in Tongguan Gold Area, Central China. *Minerals* **2020**, *10*, 89. [CrossRef]
68. Asaka, M.; Holt, R.M. Anisotropic Wellbore Stability Analysis: Impact on Failure Prediction. *Rock Mech. Rock Eng.* **2021**, *54*, 583–605. [CrossRef]
69. Wang, Y.; Liu, X.; Liang, L.; Xiong, J. Experimental study on the damage of organic-rich shale during water-shale interaction. *J. Nat. Gas Sci. Eng.* **2020**, *74*, 103103. [CrossRef]
70. Allawi, R.H. Chemical and mechanical model to analysis wellbore stability. *Pet. Sci. Technol.* **2024**, *42*, 3062–3084. [CrossRef]
71. Hu, Q.; Gu, Y.; Liu, Z.; Zeng, J.; Zhang, W.; He, L.; Li, M. Research on the Harmless Treatment of Weak Muddy Intercalation in Red-Bedded Soft Rock via Microwave Energy. *Minerals* **2023**, *13*, 1327. [CrossRef]
72. De, S.; Aastha; Sengupta, D. Depositional environment and geomechanical properties of Cambay Shale: Potential reservoir for shale oil and gas in India. *Arab. J. Geosci.* **2020**, *13*, 455. [CrossRef]

73. AlArfaj, M.; Alqahtani, H.; Alsubaie, T.; Alotaibi, M. Development of a New Inhibitive Fluid to Mitigate Shale Drilling Problems. In Proceedings of the Middle East Oil, Gas and Geosciences Show, Manama, Bahrain, 19–21 February 2023; p. D031S100R003.
74. Hu, L. A review of mechanical mechanism and prediction of natural fracture in shale. *Arab. J. Geosci.* **2022**, *15*, 474. [CrossRef]
75. Zhang, H.; Zhong, Y.; She, J.; Kuang, J. Experimental study of nano-drilling fluid based on nano temporary plugging technology and its application mechanism in shale drilling. *Appl. Nanosci.* **2019**, *9*, 1637–1648. [CrossRef]
76. Ao, T.; Yang, L.; Xie, C.; Jiang, G.; Wang, G.; Liu, Z.; He, X. Zwitterionic Silica-Based Hybrid Nanoparticles for Filtration Control in Oil Drilling Conditions. *ACS Appl. Nano Mater.* **2021**, *4*, 11052–11062. [CrossRef]
77. Ibrahim, M.A.; Saleh, T.A. Synthesis of efficient stable dendrimer-modified carbon for cleaner drilling shale inhibition. *J. Environ. Chem. Eng.* **2021**, *9*, 104792. [CrossRef]
78. Ni, X.; Shi, H.; Zhang, J.; Liu, R.; Wang, J.; Cheng, R. Modified Laponite synthesized with special wettability as a multifunctional additive in oil-based drilling fluids. *J. Pet. Sci. Eng.* **2023**, *220*, 111211. [CrossRef]
79. Aftab, A.; Ismail, A.R.; Ibupoto, Z.H. Enhancing the rheological properties and shale inhibition behavior of water-based mud using nanosilica, multi-walled carbon nanotube, and graphene nanoplatelet. *Egypt. J. Pet.* **2017**, *26*, 291–299. [CrossRef]
80. Metin, C.; Bonnecaze, R.T.; Nguyen, Q.P. The Viscosity of Silica Nanoparticle Dispersions in Permeable Media. *SPE Reserv. Eval. Eng.* **2013**, *16*, 327–332. [CrossRef]
81. Sudharsan, J.; Khare, S.K. Shale swelling inhibition by nanoparticle additives in drilling fluids—A comprehensive review of nanoparticle induced oil and gas wellbore stability. *AIP Conf. Proc.* **2023**, *2943*, 050001. [CrossRef]
82. Yang, X.; Cai, J.; Jiang, G.; Xie, J.; Shi, Y.; Chen, S.; Yue, Y.; Yu, L.; He, Y.; Xie, K. Nanoparticle plugging prediction of shale pores: A numerical and experimental study. *Energy* **2020**, *208*, 118337. [CrossRef]
83. Akilu, S.; Padmanabhan, E.; Sun, Z. A review of transport mechanisms and models for unconventional tight shale gas reservoir systems. *Int. J. Heat Mass Transf.* **2021**, *175*, 121125. [CrossRef]
84. Sheng, M.A.O.; Li, G.; Tian, S.; Huang, Z.; Chen, L. A fractal permeability model for shale matrix with multi-scale porous structure. *Fractals* **2016**, *24*, 1650002. [CrossRef]
85. Sudharsan, J.; Khare, S.K. Pore pressure transmission test as measure of shale inhibition performance of nanoparticle additives in water based drilling fluids—A comprehensive review. *Mater. Today Proc.* **2023**. [CrossRef]
86. Xia, P.; Pan, Y. Effects of nanosilica on the properties of brine-base drilling fluid. *Sci. Rep.* **2023**, *13*, 20462. [CrossRef]
87. Liu, Y.; Dai, C.; Wang, K.; Zou, C.; Gao, M.; Fang, Y.; Zhao, M.; Wu, Y.; You, Q. Study on a Novel Cross-Linked Polymer Gel Strengthened with Silica Nanoparticles. *Energy Fuels* **2017**, *31*, 9152–9161. [CrossRef]
88. Bardhan, A.; Vats, S.; Prajapati, D.K.; Halari, D.; Sharma, S.; Saxena, A. Utilization of mesoporous nano-silica as high-temperature water-based drilling fluids additive: Insights into the fluid loss reduction and shale stabilization potential. *Geoenergy Sci. Eng.* **2024**, *232*, 212436. [CrossRef]
89. Cai, J.; Gu, S.; Wang, F.; Yang, X.; Yue, Y.; Wu, X.; Chixotkin, V.F. Decreasing Coalbed Methane Formation Damage Using Microfoamed Drilling Fluid Stabilized by Silica Nanoparticles. *J. Nanomater.* **2016**, *2016*, 9037532. [CrossRef]
90. Li, Q.; Li, Z.; Zhang, H.; Sun, H.; Li, L.; Li, X.; Wang, Y. Study on Foam Drilling Fluid Stabilized with Nanomaterials Optimized with RSM. *Drill. Fluid Complet. Fluid* **2020**, *37*, 23–28. [CrossRef]
91. Kang, Y.; She, J.; Zhang, H.; You, L.; Song, M. Strengthening shale wellbore with silica nanoparticles drilling fluid. *Petroleum* **2016**, *2*, 189–195. [CrossRef]
92. Ni, X.; Jiang, G.; Li, Y.; Yang, L.; Li, W.; Wang, K.; Deng, Z. Synthesis of superhydrophobic nanofluids as shale inhibitor and study of the inhibition mechanism. *Appl. Surf. Sci.* **2019**, *484*, 957–965. [CrossRef]
93. Geng, Y.; Sun, J.; Xie, S.; Huang, X.; Wang, R.; Wang, J.; Wang, Q. Novel Use of a Superhydrophobic Nanosilica Performing Wettability Alteration and Plugging in Water-Based Drilling Fluids for Wellbore Strengthening. *Energy Fuels* **2022**, *36*, 6144–6158. [CrossRef]
94. Li, Y.; Xia, C.; Liu, X. Water-based drilling fluids containing hydrophobic nanoparticles for minimizing shale hydration and formation damage. *Heliyon* **2023**, *9*, e22990. [CrossRef]
95. Liu, F.; Zhang, C.; Li, X.; Zhang, Z.; Wang, X.; Dai, X.; Zhou, M.; Liu, Q. Investigation of the inhibition mechanism of polymer/nano-silica composite as shale inhibitor in water-based drilling fluids. *Colloids Surf. A Physicochem. Eng. Asp.* **2022**, *636*, 128099. [CrossRef]
96. Lai, N.; Fan, W.; Zhang, X.; Liu, L.; Zhou, X.; Chen, S. Temperature-sensitive polymer based nano-SiO2 composite multi-component synergistic improvement of shale stability in water-based drilling fluids. *Geoenergy Sci. Eng.* **2023**, *224*, 211498. [CrossRef]
97. Akhtarmanesh, S.; Shahrabi, M.J.A.; Atashnezhad, A. Improvement of wellbore stability in shale using nanoparticles. *J. Pet. Sci. Eng.* **2013**, *112*, 290–295. [CrossRef]
98. Huang, X.; Sun, J.; Li, H.; Wang, R.; Lv, K.; Li, H. Fabrication of a Hydrophobic Hierarchical Surface on Shale Using Modified Nano-SiO2 for Strengthening the Wellbore Wall in Drilling Engineering. *Engineering* **2022**, *11*, 101–110. [CrossRef]
99. Ospanov, Y.K.; Kudaikulova, G.A. A comprehensive review of carbon nanomaterials in the drilling industry. *J. Polym. Sci.* **2024**. [CrossRef]
100. Kosynkin, D.V.; Ceriotti, G.; Wilson, K.C.; Lomeda, J.R.; Scorsone, J.T.; Patel, A.D.; Friedheim, J.E.; Tour, J.M. Graphene Oxide as a High-Performance Fluid-Loss-Control Additive in Water-Based Drilling Fluids. *ACS Appl. Mater. Interfaces* **2012**, *4*, 222–227. [CrossRef]

101. Rafieefar, A.; Sharif, F.; Hashemi, A.; Bazargan, A.M. Rheological Behavior and Filtration of Water-Based Drilling Fluids Containing Graphene Oxide: Experimental Measurement, Mechanistic Understanding, and Modeling. *ACS Omega* **2021**, *6*, 29905–29920. [CrossRef]
102. Rana, A.; Arfaj, M.K.; Saleh, T.A. Graphene grafted with glucopyranose as a shale swelling inhibitor in water-based drilling mud. *Appl. Clay Sci.* **2020**, *199*, 105806. [CrossRef]
103. Lv, K.; Huang, P.; Zhou, Z.; Wei, X.; Luo, Q.; Huang, Z.; Yan, H.; Jia, H. Study of Janus Amphiphilic Graphene Oxide as a High-Performance Shale Inhibitor and Its Inhibition Mechanism. *Front. Chem.* **2020**, *8*, 201. [CrossRef] [PubMed]
104. Wang, X.; Li, Q.; Xie, J.; Jin, Z.; Wang, J.; Li, Y.; Jiang, K.; Fan, S. Fabrication of Ultralong and Electrically Uniform Single-Walled Carbon Nanotubes on Clean Substrates. *Nano Lett.* **2009**, *9*, 3137–3141. [CrossRef] [PubMed]
105. Okoro, E.E.; Zuokumor, A.A.; Okafor, I.S.; Igwilo, K.C.; Orodu, K.B. Determining the optimum concentration of multiwalled carbon nanotubes as filtrate loss additive in field-applicable mud systems. *J. Pet. Explor. Prod. Technol.* **2020**, *10*, 429–438. [CrossRef]
106. Liu, J.-P.; Zhang, X.-F.; Zhang, W.-C.; Lv, K.-H.; Bai, Y.-R.; Wang, J.-T.; Huang, X.-B.; Jin, J.-F.; Sun, J.-S. Carbon nanotube enhanced water-based drilling fluid for high temperature and high salinity deep resource development. *Pet. Sci.* **2022**, *19*, 916–926. [CrossRef]
107. Rana, A.; Khan, I.; Ali, S.; Saleh, T.A.; Khan, S.A. Controlling Shale Swelling and Fluid Loss Properties of Water-Based Drilling Mud via Ultrasonic Impregnated SWCNTs/PVP Nanocomposites. *Energy Fuels* **2020**, *34*, 9515–9523. [CrossRef]
108. Mahdavinezhad, M.; Kazemi-Beydokhti, A.; Sanati, A.; Malayeri, M.R. Surface Modification of Functionalized Multi-walled Carbon Nanotubes for EOR Using Low Salinity Seawater. *J. Pet. Res.* **2021**, *31*, 3–16. [CrossRef]
109. Mohd Saparti, M.D.K.; Rohani, R.; Wan Sulaiman, W.R.; Jamaluddin, N.; Khairul Zaman, N. Amine-functionalized graphene oxide in polyamine based drilling fluids for nanosized pore plug filter cake. *Geoenergy Sci. Eng.* **2023**, *230*, 212146. [CrossRef]
110. Ma, L.; Luo, P.; He, Y.; Zhang, L.; Fan, Y.; Jiang, Z. Improving the stability of multi-walled carbon nano-tubes in extremely environments: Applications as nano-plugging additives in drilling fluids. *J. Nat. Gas Sci. Eng.* **2020**, *74*, 103082. [CrossRef]
111. Hao, W.; Flynn, S.L.; Kashiwabara, T.; Alam, M.S.; Bandara, S.; Swaren, L.; Robbins, L.J.; Alessi, D.S.; Konhauser, K.O. The impact of ionic strength on the proton reactivity of clay minerals. *Chem. Geol.* **2019**, *529*, 119294. [CrossRef]
112. Pourkhalil, H.; Nakhaee, A. Effect of Nano ZnO on wellbore stability in shale: An experimental investigation. *J. Pet. Sci. Eng.* **2019**, *173*, 880–888. [CrossRef]
113. Dejtaradon, P.; Hamidi, H.; Chuks, M.H.; Wilkinson, D.; Rafati, R. Impact of ZnO and CuO nanoparticles on the rheological and filtration properties of water-based drilling fluid. *Colloids Surf. A Physicochem. Eng. Asp.* **2019**, *570*, 354–367. [CrossRef]
114. Bayat, A.E.; Shams, R. Appraising the impacts of SiO_2, ZnO and TiO_2 nanoparticles on rheological properties and shale inhibition of water-based drilling muds. *Colloids Surf. A Physicochem. Eng. Asp.* **2019**, *581*, 123792. [CrossRef]
115. Srivastava, V.; Beg, M.; Sharma, S.; Choubey, A.K. Application of manganese oxide nanoparticles synthesized via green route for improved performance of water-based drilling fluids. *Appl. Nanosci.* **2021**, *11*, 2247–2260. [CrossRef]
116. Ejtemaee, P.; Khamehchi, E. Experimental investigation of rheological properties and formation damage of water-based drilling fluids in the presence of Al_2O_3, Fe_3O_4, and TiO_2 nanoparticles. *Biointerface Res. Appl. Chem.* **2020**, *10*, 5886–5894. [CrossRef]
117. Jiang, G.; Dong, T.; Cui, K.; He, Y.; Quan, X.; Yang, L.; Fu, Y. Research status and development directions of intelligent drilling fluid technologies. *Pet. Explor. Dev.* **2022**, *49*, 660–670. [CrossRef]
118. Alvi, M.A.; Belayneh, M.; Bandyopadhyay, S.; Minde, M.W. Effect of Iron Oxide Nanoparticles on the Properties of Water-Based Drilling Fluids. *Energies* **2020**, *13*, 6718. [CrossRef]
119. Al-Shargabi, M.; Davoodi, S.; Wood, D.A.; Al-Musai, A.; Rukavishnikov, V.S.; Minaev, K.M. Nanoparticle applications as beneficial oil and gas drilling fluid additives: A review. *J. Mol. Liq.* **2022**, *352*, 118725. [CrossRef]
120. Onaizi, S.A. Mineral oil-based drilling fluid formulation using biosurfactant and nanoparticles with good rheological behavior and excellent H2S scavenging capability. *JCIS Open* **2024**, *13*, 100100. [CrossRef]
121. Ahasan, M.H.; Alahi Alvi, M.F.; Ahmed, N.; Alam, M.S. An investigation of the effects of synthesized zinc oxide nanoparticles on the properties of water-based drilling fluid. *Pet. Res.* **2022**, *7*, 131–137. [CrossRef]
122. Aftab, A.; Ali, M.; Sahito, M.F.; Mohanty, U.S.; Jha, N.K.; Akhondzadeh, H.; Azhar, M.R.; Ismail, A.R.; Keshavarz, A.; Iglauer, S. Environmental Friendliness and High Performance of Multifunctional Tween 80/ZnO-Nanoparticles-Added Water-Based Drilling Fluid: An Experimental Approach. *ACS Sustain. Chem. Eng.* **2020**, *8*, 11224–11243. [CrossRef]
123. Medhi, S.; Gupta, D.K.; Sangwai, J.S. Impact of zinc oxide nanoparticles on the rheological and fluid-loss properties, and the hydraulic performance of non-damaging drilling fluid. *J. Nat. Gas Sci. Eng.* **2021**, *88*, 103834. [CrossRef]
124. Alsaba, M.; Al Marshad, A.; Abbas, A.; Abdulkareem, T.; Al-Shammary, A.; Al-Ajmi, M.; Kebeish, E. Laboratory evaluation to assess the effectiveness of inhibitive nano-water-based drilling fluids for Zubair shale formation. *J. Pet. Explor. Prod. Technol.* **2020**, *10*, 419–428. [CrossRef]
125. Li, Q.; Berraud-Pache, R.; Souprayen, C.; Jaber, M. Intercalation of lecithin into bentonite: pH dependence and intercalation mechanism. *Appl. Clay Sci.* **2023**, *244*, 107079. [CrossRef]
126. Song, K.; Wu, Q.; Li, M.; Ren, S.; Dong, L.; Zhang, X.; Lei, T.; Kojima, Y. Water-based bentonite drilling fluids modified by novel biopolymer for minimizing fluid loss and formation damage. *Colloids Surf. A Physicochem. Eng. Asp.* **2016**, *507*, 58–66. [CrossRef]
127. Abdou, M.I.; Al-sabagh, A.M.; Dardir, M.M. Evaluation of Egyptian bentonite and nano-bentonite as drilling mud. *Egypt. J. Pet.* **2013**, *22*, 53–59. [CrossRef]

128. Falode, O.A.; Ehinola, O.A.; Nebeife, P.C. Evaluation of local bentonitic clay as oil well drilling fluids in Nigeria. *Appl. Clay Sci.* **2008**, *39*, 19–27. [CrossRef]
129. İşçi, E.; Turutoğlu, S.İ. Stabilization of the mixture of bentonite and sepiolite as a water based drilling fluid. *J. Pet. Sci. Eng.* **2011**, *76*, 1–5. [CrossRef]
130. Li, J.; Qiu, Z.; Zhong, H.; Zhao, X.; Liu, Z.; Huang, W. Effects of water-based drilling fluid on properties of mud cake and wellbore stability. *J. Pet. Sci. Eng.* **2022**, *208*, 109704. [CrossRef]
131. van Vliet, J.P.M.; Hjh Rosita, H.H. Reservoir Impairment Mitigation: Successful Use of Graded Carbonate Drilling Fluids. In Proceedings of the SPE Asia Pacific Oil and Gas Conference and Exhibition, Jakarta, Indonesia, 20–22 April 1999; p. SPE–54284-MS.
132. Quintero, L.; Jones, T.A. An Alternative Drill-in Fluid System for Low-Pressure Reservoirs. In Proceedings of the SPE European Formation Damage Conference, The Hague, The Netherlands, 13–14 May 2003; p. SPE–82280-MS.
133. Basfar, S.; Elkatatny, S. Micronized calcium carbonate to enhance water-based drilling fluid properties. *Sci. Rep.* **2023**, *13*, 18295. [CrossRef]
134. Alkhalaf, S.; Alawami, M.; Wagle, V.; Al-Yami, A. Less Damaging Drilling Fluids: Development and Lab Testing. In Proceedings of the International Petroleum Technology Conference, Beijing, China, 26–28 March 2019; p. D031S062R001.
135. Dehghani, F.; Kalantariasl, A.; Saboori, R.; Sabbaghi, S.; Peyvandi, K. Performance of carbonate calcium nanoparticles as filtration loss control agent of water-based drilling fluid. *SN Appl. Sci.* **2019**, *1*, 1466. [CrossRef]
136. Li, Q. Bio-Organoclays for Oil-Based Drilling Fluids Bio-Argiles Organiques Pour Fluides de Forage à Base de Pétrole. Ph.D. Thesis, Sorbonne Université, Paris, France, 2023.
137. Borah, B.; Das, B.M. A review on applications of bio-products employed in drilling fluids to minimize environmental footprint. *Environ. Chall.* **2022**, *6*, 100411. [CrossRef]
138. Liu, J.; Zhou, F.; Deng, F.; Zhao, H.; Wei, Z.; Long, W.; Evelina, A.; Ma, C.; Chen, S.; Ma, L. Improving the rheological properties of water-based calcium bentonite drilling fluids using water-soluble polymers in high temperature applications. *J. Polym. Eng.* **2022**, *42*, 129–139. [CrossRef]
139. Li, Q.; Berraud-Pache, R.; Yang, Y.; Souprayen, C.; Jaber, M. Biocomposites based on bentonite and lecithin: An experimental approach supported by molecular dynamics. *Appl. Clay Sci.* **2023**, *231*, 106751. [CrossRef]
140. Li, Z.; Li, Q.; Yang, G.; Zhang, F.; Ma, T. The Synthesis and Application of a New Plugging Inhibitor PAS-5 in Water-Based Drilling Fluid. In Proceedings of the SPE Middle East Oil and Gas Show and Conference, Manama, Barhain, 18–21 March 2019; p. D022S052R001.
141. Huang, X.; Lv, K.; Sun, J.; Lu, Z.; Bai, Y.; Shen, H.; Wang, J. Enhancement of thermal stability of drilling fluid using laponite nanoparticles under extreme temperature conditions. *Mater. Lett.* **2019**, *248*, 146–149. [CrossRef]
142. Xiong, Z.-Q.; Li, X.-D.; Fu, F.; Li, Y.-N. Performance evaluation of laponite as a mud-making material for drilling fluids. *Pet. Sci.* **2019**, *16*, 890–900. [CrossRef]
143. Huang, X.-B.; Sun, J.-S.; Huang, Y.; Yan, B.-C.; Dong, X.-D.; Liu, F.; Wang, R. Laponite: A promising nanomaterial to formulate high-performance water-based drilling fluids. *Pet. Sci.* **2021**, *18*, 579–590. [CrossRef]
144. Huang, X.; Shen, H.; Sun, J.; Lv, K.; Liu, J.; Dong, X.; Luo, S. Nanoscale Laponite as a Potential Shale Inhibitor in Water-Based Drilling Fluid for Stabilization of Wellbore Stability and Mechanism Study. *ACS Appl. Mater. Interfaces* **2018**, *10*, 33252–33259. [CrossRef]
145. Meng, X.; Zhang, Y.; Zhou, F.; An, Q. Influence of carbon ash on the rheological properties of bentonite dispersions. *Appl. Clay Sci.* **2014**, *88–89*, 129–133. [CrossRef]
146. Oni, O.; Fadairo, A.S.; Gosnold, W.; Egenhoff, S.; Adeyemi, G.; Ling, K. Investigating the suitability of North Dakota fly ash as fluid loss reducing additive in densified water-based drilling fluid. *Powder Technol.* **2023**, *430*, 118972. [CrossRef]
147. Bageri, B.S.; Adebayo, A.R.; Al Jaberi, J.; Patil, S. Effect of perlite particles on the filtration properties of high-density barite weighted water-based drilling fluid. *Powder Technol.* **2020**, *360*, 1157–1166. [CrossRef]
148. Al Jaberi, J.; Bageri, B.; Alzaidi, G.; Alruhaili, A. Influence of perlite on the stability of filter cake in low-density drilling fluids system. *Geoenergy Sci. Eng.* **2024**, *243*, 213305. [CrossRef]
149. Zhang, B.; Wang, Q.; Du, W.; Li, Y.; Zhang, J.; Zhang, J.; Matejdes, M.; Slaný, M.; Gang, C. Multi-Mixed Metal Hydroxide as a Strong Stratigraphic Nanoclay Inhibitor in Solid-Free Drilling Fluid. *Nanomaterials* **2022**, *12*, 3863. [CrossRef] [PubMed]
150. Gokapai, V.; Pothana, P.; Ling, K. Nanoparticles in Drilling Fluids: A Review of Types, Mechanisms, Applications, and Future Prospects. *Eng* **2024**, *5*, 2462–2495. [CrossRef]
151. Zhuang, G.; Li, Q.; Zhang, Z.; Bergaya, F.; Yuan, P. Chapter 2—Application of bentonite in water-based drilling fluids. In *Clay Science in Drilling and Drilling Fluids*; Zhuang, G., Yuan, P., Eds.; Elsevier: Amsterdam, The Netherlands, 2024; Volume 11, pp. 21–53.
152. Saleh, T.A.; Nur, M.M.; Satria, M.; Al-Arfaj, A.A. Synthesis of novel hydrophobic nanocomposite-modified silica as efficient shale inhibitor in fuel industry. *Surf. Interfaces* **2023**, *38*, 102837. [CrossRef]
153. Quainoo, A.K.; Negash, B.M.; Bavoh, C.B.; Ganat, T.O.; Tackie-Otoo, B.N. A perspective on the potential application of bio-inhibitors for shale stabilization during drilling and hydraulic fracturing processes. *J. Nat. Gas Sci. Eng.* **2020**, *79*, 103380. [CrossRef]
154. Aravind, J.; Kamaraj, M. *Carbon-Based Composites and Nanocomposites: Adsorbents and Membranes for Environmental Remediation*; Walter de Gruyter GmbH & Co KG: Berlin, Germany, 2024.

155. Abdullah, M.M.; Al-Lohedan, H.A. Nanomaterials in petroleum industry. In *Handbook of Nanomaterials*; Elsevier: Amsterdam, The Netherlands, 2024; Volume 1, pp. 617–648.
156. Chen, S.; Huang, D.; Xu, Z.; Bai, Y.; Xie, G.; Li, X.; Chen, H.; Zhou, C.; Luo, P.; Feng, R. Synthesis and mechanism analysis of a non-toxic amine-based clay mineral surface hydration intercalation inhibitor. *J. Mol. Liq.* **2024**, *399*, 124362. [CrossRef]
157. Vryzas, Z.; Kelessidis, V.C.; Bowman, M.B.; Nalbantian, L.; Zaspalis, V.; Mahmoud, O.; Nasr-El-Din, H.A. Smart Magnetic Drilling Fluid With In-Situ Rheological Controllability Using Fe3O4 Nanoparticles. In Proceedings of the SPE Middle East Oil & Gas Show and Conference, Manama, Bahrain, 6–9 March 2017; p. D031S024R001.
158. Li, Q.; Zhuang, G.; Yuan, P.; Bergaya, F. Chapter 12—Future challenges related to clay minerals in drilling and drilling fluids. In *Clay Science in Drilling and Drilling Fluids*; Zhuang, G., Yuan, P., Eds.; Elsevier: Amsterdam, The Netherlands, 2024; Volume 11, pp. 313–338.
159. Cui, K.; Jiang, G.; Xie, C.; Yang, L.; He, Y.; Shen, X.; Wang, X. A novel temperature-sensitive expandable lost circulation material based on shape memory epoxy foams to prevent losses in geothermal drilling. *Geothermics* **2021**, *95*, 102145. [CrossRef]
160. Yang, L.; Xie, C.; Ao, T.; Cui, K.; Jiang, G.; Bai, B.; Zhang, Y.; Yang, J.; Wang, X.; Tian, W. Comprehensive evaluation of self-healing polyampholyte gel particles for the severe leakoff control of drilling fluids. *J. Pet. Sci. Eng.* **2022**, *212*, 110249. [CrossRef]

minerals

MDPI

Review

Application of Nanofibrous Clay Minerals in Water-Based Drilling Fluids: Principles, Methods, and Challenges

Guanzheng Zhuang [1,*], Jiajun Zhang [1], Jinrong Chen [1], Qian Liu [1], Wenxiao Fan [1] and Qiang Li [2]

1 School of Environmental Science and Engineering, Guangdong University of Technology, Guangzhou 510006, China; 2112207019@mail2.gdut.edu.cn (J.Z.); 2112207114@mail2.gdut.edu.cn (J.C.); 2112307059@mail2.gdut.edu.cn (Q.L.); fanwenxiao@gdut.edu.cn (W.F.)

2 Faculty of Petroleum, China University of Petroleum-Beijing at Karamay, Karamay 834000, China; liqiang2024@cupk.edu.cn

* Correspondence: zhuangguanzheng@gdut.edu.cn

Abstract: Nanofibrous clay minerals, specifically palygorskite (Pal) and sepiolite (Sep), have been becoming a new generation of rheological additives for drilling fluid systems due to their unique nanostructure, high performance, environmentally benign nature, and cost-effectiveness. These nanoclay minerals exhibit excellent colloidal and rheological properties in aqueous systems, even in saline and high-temperature environments. Although Pal and Sep have been employed as auxiliary rheological additives in a few cases to enhance the salt resistance of conventional water-based drilling fluids (WBDFs), these two clay minerals have not yet been used on a large scale due to a lack of understanding of their structures and properties, as well as the control of their performance. This paper presents a comprehensive review of the clay mineralogy, colloidal chemistry, rheological behaviors, and filtration properties of nanofibrous clay minerals in WBDFs, with critical comments. It also discusses the challenges and prospects for further research. This review provides new insights into fundamental and applied studies of nanofibrous clay minerals and helps promote the large-scale application of nanofibrous clay products in drilling fluids.

Keywords: palygorskite; sepiolite; drilling muds; rheological additives; colloidal chemistry

1. Introduction

Currently, oil and gas are the most and third-most prominent energy sources, respectively. The International Energy Agency (IEA) has forecast that oil and gas consumption demand will continue to rise until 2050. Consequently, the extraction of additional oil and gas represents one of the most urgent tasks which must be addressed before the advent of the new energy era. However, extracting oil and gas from the earth is becoming increasingly challenging. Drilling operations must overcome a number of challenges, including high temperatures, high pressure, unstable shales, and salty formations [1]. It is therefore evident that there is a need for high-performance drilling fluids in harsh environments.

Drilling fluids, also called drilling muds, are indispensable for the successful execution of drilling operations. The principal functions of drilling fluids are to suspend and transport drilling cuttings, clean the borehole, stabilize the wellbore by forming filter cakes, and balance the formation pressure [2,3]. In general, drilling fluids are classified into two categories: water-based drilling fluids (WBDFs) and oil-based drilling fluids (OBDFs). WBDFs are typically composed of fresh water, bentonite, and other additives [4]. OBDFs are generally formulated with base oils (principally diesel, low-toxicity mineral oils, and synthetic oils), organoclays, and other additives [5]. OBDFs are known for their excellent shale-inhibiting properties, lubricity, and salt resistance [5]. Nevertheless, they are frequently criticized for their high cost and toxicity derived from base oils. In contrast, WBDFs are inexpensive and environmentally friendly. Therefore, WBDFs are widely used, especially in challenging drilling operations and environmentally sensitive locations.

Despite the considerable advantages in terms of cost and environmental care offered by WBDFs, their application still encounters significant obstacles, particularly the difficulty of rheological control. The rheology of drilling fluids, including properties such as the viscosity, yield point, and gel strength, is a critical factor in the efficiency and safety of drilling operations in the oil and gas industry. Drilling fluids usually exhibit non-Newtonian behavior, with shear-thinning characteristics where the viscosity decreases with an increasing shear rate. This rheological behavior is essential for facilitating effective hole cleaning, as it allows the fluid to flow easily near the drill bit while maintaining sufficient viscosity to suspend and transport cuttings. The yield point and gel strength of the fluid further contribute to this by ensuring that cuttings are suspended even at low flow rates and that they do not settle when circulation stops. These properties also play a significant role in pressure control by influencing the equivalent circulating density (ECD), which helps balance hydrostatic pressure to prevent both formation fluid influx and formation fracturing. Consequently, the proper management of drilling fluid rheology is vital for maintaining wellbore stability, minimizing operational issues, and optimizing overall drilling efficiency.

The rheological behaviors of WBDFs are generally governed by bentonite, a natural clay with montmorillonite (Mt) as the primary component [6–12]. Mt can form a network of interconnected particles, namely a "house of cards" structure when it swells and even exfoliates in water. This property endows it with excellent rheological properties in aqueous systems, making it an invaluable component in WBDFs [6,13–21]. However, the properties of Mt-based drilling fluids are severely compromised in saline and high-temperature formations due to the large surface charge density and inadequate stability of Mt [6,18–20,22–25]. While water-soluble polymers, such as xanthan gum and carboxymethyl cellulose, can enhance specific aspects of fluid rheology, such as the low-shear viscosity, fluid loss control, and salt resistance, they are still limited due to several key factors related to performance, cost, and their functional properties. For example, the rheology of polymers is sensitive to temperature changes, especially at low and high temperatures. Therefore, it is urgent to develop new rheological additives for WBDFs with excellent rheological properties, thermal stability, and cost-effectiveness.

In recent years, plenty of new materials have been developed to improve the applied properties of WBDFs. These include improvements to the rheological properties under saline and high-temperature conditions, shale-inhibiting ability, and filtration properties. Nanomaterials are particularly interesting among these additives [26–28]. As a consequence, a novel concept, designated "nano-based drilling fluids" (also referred to as "nanomaterials-based drilling fluids" or "nano-enhanced drilling fluids"), is emerging [29–31]. Nanomaterials which have been reported for use in WBDFs include graphene nanoplatelets [32,33], nanocomposites [34,35], and nanoclay minerals [36–46]. Nanoclay minerals represent a promising class of additives for WBDFs, offering several advantages, such as their natural origin, environmental friendliness, low cost, and excellent rheological properties.

Nanofibrous clay minerals, specifically palygorskite (Pal) and sepiolite (Sep), are some of the most important rheological additives for high-performance WBDFs. Pal and Sep are unique fibrous clay minerals with distinct structures and properties which make them valuable in various industrial applications. Both minerals have the characteristic of a 2:1 layered structure. However, their fibrous morphology distinguishes them from other 2:1 clay minerals like Mt or illite. Pal has a needle-like structure consisting of ribbons of silica tetrahedra linked by magnesium, aluminum, or iron octahedra, creating channels along its fibers to accommodate water molecules and other ions. Similarly, Sep has a fibrous structure but with a more open, lath-like configuration, allowing for an even higher specific surface area and porosity. These structural characteristics impart several important properties; both minerals exhibit a high specific surface area, strong sorption capacities, and significant thermal stability.

Additionally, their fibrous nature provides excellent rheological properties, particularly in suspensions, which contribute to increased viscosity and gel formation. The ability of Pal and Sep to absorb water and other fluids and their chemical stability make them particularly useful in drilling fluids, where they enhance the fluid's rheology. They exhibit excellent colloidal and rheological properties, even in saline and high-temperature environments [47]. These two clay minerals also offer notable advantages over other synthetic nanomaterials because they are naturally occurring, cost-effective, and environmentally friendly [48]. Indeed, Pal and Sep have been used in drilling fluids in some cases. They are included in the specifications of the American Petroleum Institute (API) standard. Neaman and Singer [49] conducted a study investigating the rheological properties of three Pal samples sourced from Sacalum (Mexico), Florida (USA), and Georgia (USA). Their findings indicated that the WBDFs with 5% (w/v) Pal exhibited an apparent viscosity (AV) of 16.63–23.97 mPa·s and a yield point (YP) of 13.06–16.39 Pa (at a shear rate of 1000 s^{-1}), which satisfied the API requirements. A recent study [42] demonstrated that Turkish Sep outperformed Wyoming bentonite in WBDFs. The remarkable thickening ability of Pal and Sep in WBDFs contributes to their capacity to transport drill cuttings and suspend weight agents. Aside from this, WBDFs prepared with Pal or Sep also exhibit shear-thinning behavior [40,50–53], which is highly beneficial for improving the drilling rate and cutting carrying ability. Recent studies have demonstrated that Pal and Sep fluids presented thixotropy [52], with even more pronounced effects than those observed in Wyoming bentonite-based WBDFs [40]. It is postulated that Pal and Sep may be invaluable in overcoming drilling challenges, especially in high-temperature and high-pressure formations, shale formations, and saline formations [54].

Although nanofibrous clay minerals exhibit promising rheological properties in WBDFs, their large-scale application is still constrained by both theoretical and engineering challenges. The use of these clay minerals in drilling fluids is a multidisciplinary issue which requires an interdisciplinary approach, incorporating insights from clay mineralogy, colloid chemistry, rheology, and filtrate control. Therefore, a comprehensive and critical understanding of the current research is essential to clarify fundamental scientific issues, performance regulation, challenges, and future directions for the application of nanofibrous clay minerals in WBDFs. The primary goal of this review is to comprehensively assess the current state of research on the application of nanofibrous clay minerals, specifically Pal and Sep, in WBDFs. With the increasing demand for more efficient and environmentally friendly drilling operations, it is crucial to understand how these unique clay minerals enhance the rheological, thermal, and filtration properties of drilling fluids. This paper aims to synthesize existing knowledge, identify the mechanisms by which Pal and Sep influence drilling fluid performance, and highlight their potential advantages over traditional additives. Moreover, this review will explore the challenges and limitations associated with their usage, providing insights into future research directions which could further optimize their application in the oil and gas industry.

2. Clay Mineralogy

2.1. Crystal Structure

Pal and Sep exhibit a layered structure comprising an octahedral sheet sandwiched by two tetrahedral sheets. The tetrahedral basal oxygen atoms are arranged in a continuous plane at a distance of approximately 6.6 Å. However, in contrast to the ideal 2:1 clay mineral, the apical oxygen atoms of the tetrahedra point away from the basal oxygen atom plane in opposing directions to form ribbons of joined pyroxene-like chains [55]. The apical oxygen atoms of the tetrahedra partially contribute to the formation of the coordination unit of the octahedral sheet. Consequently, there is a region of the structure where channels form adjacent to the basal oxygen planes between two 2:1 layers. The dimensions of these channels are 3.7 × 6.4 Å for Pal and 3.7 × 10.6 Å for Sep (Figure 1) [56]. The key features of Pal and Sep are as follows: (1) continuous tetrahedral basal oxygen planes, (2) an inverted tetrahedral arrangement which forms ribbons of joined pyroxene-like chains, and (3) a discontinuous octahedral sheet. Due to the distinctive structural attributes of Pal and Sep, they are generally regarded as a separate group (i.e., the palygorskite-sepiolite group).

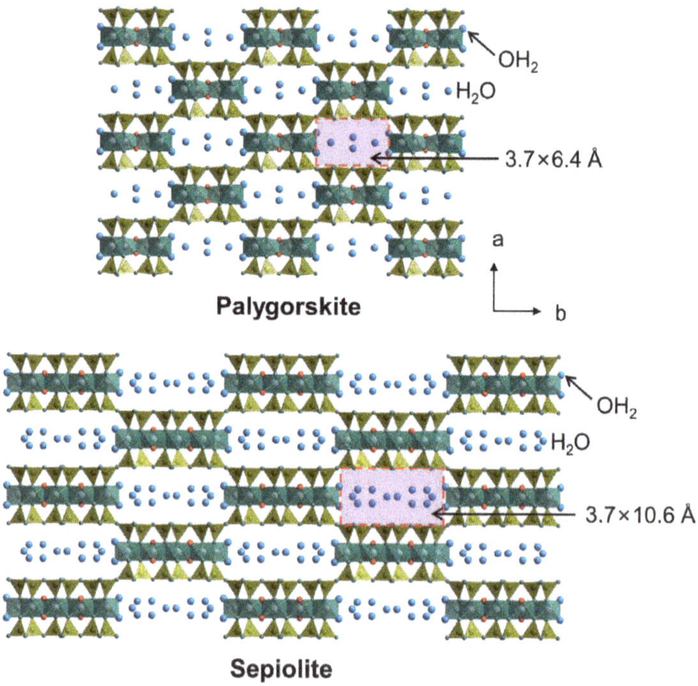

Figure 1. Structures of Pal and Sep.

2.2. Chemical Composition

The chemical formula of Pal is $(Mg_{5-y-z}R^{3+}{}_{y}\square_{z})(Si_{8-x}R^{3+}{}_{x})O_{20}(OH)_2(OH_2)_4\cdot R^{2+}{}_{\frac{x-y+2z}{2}}(H_2O)_4$, and that of Sep is $(Mg_{8-y-z}R^{3+}{}_{y}\square_{z})(Si_{12-x}R^{3+}{}_{x})O_{30}(OH)_4(OH_2)_4\cdot R^{2+}{}_{\frac{x-y+2z}{2}}(H_2O)_8$. Si^{4+} ions predominantly occupy the centers of the tetrahedra, while the octahedral centers are primarily composed of Mg^{2+} and Al^{3+} ions. Therefore, Pal and Sep can be classified as Mg-rich aluminum silicate minerals. It should be noted, however, that other cations can replace the cations in the tetrahedra and octahedra. For example, Si^{4+} ions in the tetrahedra can be substituted with Al^{3+} ions, and the Al^{3+} ions in the octahedra can be replaced by Mg^{2+} and Fe^{3+} ions. In addition, there are also vacancies and different forms of water present in these two clay minerals' structures, including hydroxyl (OH), coordinated water (OH_2), and zeolitic water (H_2O). For Pal, there is a general consensus that the OH groups are part of the octahedral anion coordination of the M1 and M2 sites, which occur well within the octahedral strips. The OH_2 is part of the coordination unit around the M3 site along the edges of the octahedral strips, where the two hydrogen atoms are required for charge balance. In the case of Sep, the OH groups are part of the inner octahedral strip coordinating with M1, M2, and M3, whereas the OH_2 groups are along the edges of the octahedral strip coordinating with M4, and this pattern is similar to that of Pal. H_2O molecules occupy the channels of Pal and Sep, and they can be removed at low temperatures (e.g., below 200 °C).

2.3. Morphology

Pal and Sep are in a group of fibrous clay minerals which exhibit elongated and needle-like morphological features. Scanning and transmission electron microscopy (SEM and TEM) observations revealed that Pal and Sep are fibrous (Figure 2), with fibers developing along the c axis. These fibers, sometimes referred to as rods, are composed of several laths, which are the crystals of Pal and Sep. Aside from this, the laths and fibers often aggregate into bundles, forming a fibrous filamentous structure at the macroscopic level. Although Pal and Sep show fibrous morphological characteristics, samples from different deposits exhibit significant variations in their detailed features. These variations include the fiber length, width, thickness, and aggregation arrangements. Generally, Pal and Sep fibers have lengths of 0.2–2 μm, widths of 10–30 nm, and thicknesses of 5–10 nm [47,56,57].

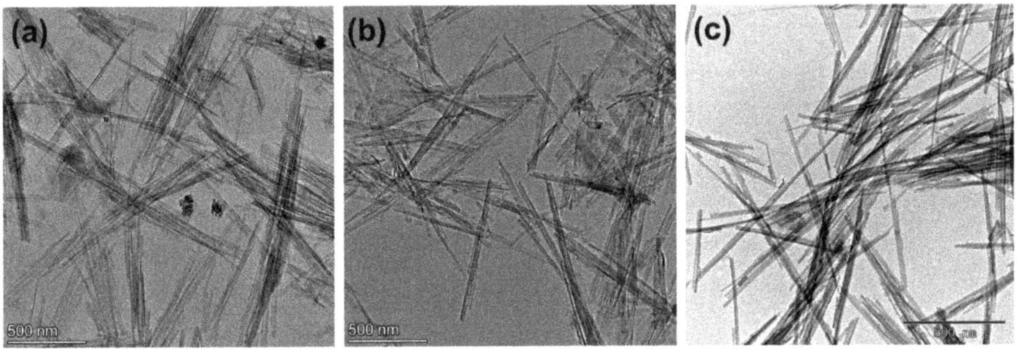

Figure 2. TEM images of (**a**) Pal from Georgia, USA, (**b**) Pal from Mingguang, China, and (**c**) Sep from Spain.

2.4. Properties and Applications

Pal and Sep can be classified into two principal categories, namely colloidal and non-colloidal applications. Non-colloidal applications are primarily composed of fillers and adsorbents, which are mainly evaluated based on their color and adsorption capacity. Pal and Sep are mainly composed of O, Si, Al, and Mg, resulting in a predominantly white-gray color. However, they may exhibit a yellow-red hue when the Fe content is elevated. The color of Pal and Sep usually only affects the appearance of products, with no significant effect on their application properties. The adsorption capacity of Pal and Sep is primarily determined by their specific surface areas and pore volumes. Due to their nanofibrous morphology and abundance of micropores, Pal and Sep exhibit high specific surface areas, generally with values of 150 m^2/g and 300 m^2/g, respectively. The colloidal applications of Pal and Sep are based on their surface hydrophilicity and nanofibrous morphology. The numerous hydrophilic groups on the surface of Pal and Sep, including silanol and aluminol groups, facilitate the dispersion of clay mineral nanofibers in water, forming a haystack structure. This structure gives Pal and Sep remarkable rheological properties [47,58–61].

3. Rheological Properties of Pal and Sep in WBDFs

3.1. Network Formation Mechanism

The network formation mechanism of fibrous clay minerals in water is primarily driven by their distinctive crystal structure and surface chemistry. These minerals consist of elongated, needle-like fibers with high aspect ratios, which enable them to form intricate, three-dimensional networks when dispersed in an aqueous medium. Upon hydration, the surfaces of these fibers become charged due to the isomorphic substitution of cations and the dissociation of surface hydroxyl groups, leading to the development of an electrostatic double layer [53,62]. This charged surface facilitates repulsion among the fibers, preventing aggregation and promoting a stable suspension. As the concentration of these clay minerals increases in water, the fibers interact through van der Waals forces and hydrogen bonding, creating a physically entangled network. This network is further stabilized by the high specific surface area and the presence of exchangeable cations, which can bridge the fibers and enhance the network's connectivity. The resulting structure significantly influences the rheological properties of the dispersion, imparting high viscosity and yield stress, which are advantageous for various applications, including WBDFs. Factors such as the pH level, ionic strength, and the presence of other additives can modify interparticle interactions and, consequently, the dispersion's rheological behavior. Additionally, the fibrous network exhibits thixotropic behavior [40,42,52], where the structure aligns and disentangles under shear, reducing viscosity, and then reassembles when the shear is removed, ensuring stability and flowability. These properties make Pal and Sep highly effective in stabilizing suspensions and controlling the rheology of WBDFs, enhancing their utility in high-performance drilling operations.

However, the mechanism of network formation of Pal and Sep in an aqueous system is still unclear. On the one hand, conventional SEM or TEM cannot observe the real structure of aqueous samples in situ. Although cryo-SEM offers a novel approach to studying the microstructure of liquid samples, the results do not fully reflect the actual clay mineral dispersions. On the other hand, the formation mechanism of the fibrous clay mineral network structure is mainly based on an aqueous system without electrolytes. The network structure of these clay minerals under different electrolyte conditions is still unclear. Since a drilling fluid is a complex multi-phase dispersion system, understanding the network structure of the minerals themselves and the influence of other additives or contaminants (particularly electrolytes) is essential for an appreciation of the rheological properties of drilling fluids.

3.2. *Influence of Clay Concentration*

The concentration of Pal or Sep significantly influences the rheological behavior of water-based drilling fluids due to their unique fibrous structure and ability to form interconnected networks within the fluid. As the concentration of these nanofibrous clay minerals increases, the degree of network formation intensifies, leading to a more pronounced impact on the fluid's viscosity and yield stress. At low concentrations, the number of nanofibers is insufficient to form a network structure [63] (Figure 3). Thus, the limited number of nanofibers does not apparently affect the flow behavior of drilling fluids. For example, the Pal fluid exhibits nearly Newtonian flow at a lower concentration of 1% (w/v), but the flow becomes pseudoplastic as the suspension concentration increases [50]. The shear rate threshold at which the flow transitions from non-Newtonian to linear depends on the clay concentration. High clay concentrations result in a preference for non-Newtonian fluids. The viscosity and gel strength of Pal and Sep in WBDFs typically increase with an increasing clay concentration [50,53,63,64] due to the increased internal friction of the fluid caused by the presence of a greater number of clay mineral particles. Thus, the concentration of Pal or Sep directly correlates with the thickening and gelation properties of the drilling fluid, influencing its performance in various drilling operations. However, the concentration of nanofibrous clay minerals added to WBDFs is not fixed; it varies with the origin, mineral content, physicochemical properties, and drilling fluid formulation.

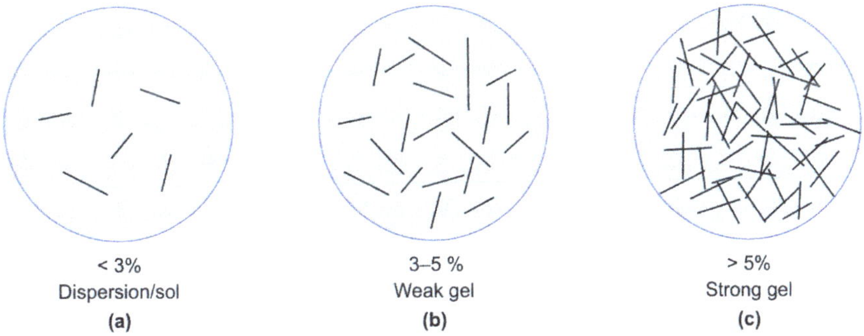

Figure 3. Influence of the fibrous clay minerals' concentration on the gel structure: (**a**) a dispersion or sol at a very low concentration, (**b**) a weak gel at a medium concentration, and (**c**) a strong gel at a high concentration.

3.3. Influence of Morphology and Fiber Arrangements

The morphology and fiber arrangements of Pal and Sep significantly influence the rheological behavior of their dispersions. Neaman and Singer [50] found that the viscosities and gel strengths of several Pal dispersions at a pH value of seven were linearly related to the ellipticity (L/W ratio) of the fibers. Later, Balter et al. demonstrated that Pal with longer fibers produced a greater viscosity in both fresh and saline water [65]. The length-to-width (L/W) ratio of the fibers is a crucial determinant of viscosity, with higher L/W ratios correlating with increased viscosity due to the greater surface area for interaction and entanglement, which facilitates a robust network structure within the fluid. This network formation, enhanced by longer fibers, increases internal friction, thereby elevating the viscosity and yield point. Effective dispersion of fiber bundles into individual fibers without breaking them is essential for maintaining high L/W ratios and achieving optimal rheological properties. Furthermore, the interactions between fibers, such as entanglement and electrostatic attractions, are more pronounced with higher L/W ratios, contributing to a stronger network. The application of higher shear rates during processing can also enhance fiber dispersion and alignment, promoting better network formation and improved rheological behavior. In summary, the rheological properties of Pal and Sep dispersions are intricately linked to their fiber morphology and arrangements, with longer, thinner fibers forming more effective networks which result in a higher viscosity and higher yield points. Proper processing techniques which preserve these morphological characteristics are essential for optimizing the performance of Pal and Sep in drilling fluids.

Since Pal and Sep nanofibers are typically found in bulk bundles, separating the fibers through special processing methods is essential. These include physical methods [66–72] and chemical methods [73–75]. For example, extrusion, slurring, surface modification, and high-pressure homogenization are effective methods for separating the fibers. Notably, excessive mechanical action may break the fibers into shorter ones, which is detrimental to the colloidal use of Pal and Sep [76]. It is therefore recommended that the disaggregation and dispersion of crystal bundles be carried out in a way that does not damage the L/W ratio. In the case of WBDFs, it was demonstrated that the dispersion of Pal or Sep at elevated shear rates resulted in enhanced rheological properties [77].

3.4. Influence of pH Conditions

The pH conditions significantly influence the rheological properties of Pal or Sep dispersions. On the one hand, the pH condition affects the flow behavior of the clay mineral dispersions. For example, Pal behaves as a nearly Newtonian fluid under basic conditions, whereas it behaves as a non-Newtonian fluid under acidic conditions [50]. On the other hand, the pH level regulates the viscosity of the fluid by affecting the association of nanofibers. A recent study [75] reported that the viscosity of Pal dispersions increased when the pH value increased from 1 to 11 but decreased when the pH value exceeded 12. Similarly, Liu et al. [53] observed that the yield stress of Sep dispersions slightly increased as the pH level increased from 2 to 8 and then decreased at pH values greater than 8. Sep dispersions typically exhibit zero yield stress at pH values above 11 [78,79]. The pH level influences the colloidal and rheological properties of Pal and Sep dispersions by altering the surface charge and fiber aggregation. The surface charge of Pal and Sep fibers is pH-dependent, affecting the electrostatic interactions between fibers. According to the literature [50,53,80,81], the point of zero charges (PZC) of Pal and Sep is usually between pH levels of 2 and 5. At a pH value equal to the PZC, the net electrostatic charge of the clay surface is zero. In this configuration, the van der Waals forces predominate the interaction between the nanofibers, resulting in the clay fibers being arranged in a parallel configuration [50]. At a pH level slightly above the PZC (e.g., approximately seven), Pal and Sep exhibit the optimal rheological properties. This is attributed to a dynamic equilibrium between the van der Waals forces and electrostatic repulsion between the nanofibers, resulting in a weak attraction between the nanofibers to form a network structure while maintaining reasonable repulsion to prevent a parallel arrangement (Figure 4). However, when the pH value is significantly greater than the PZC (e.g., above 11) or significantly smaller than the PZC, the nanofibers are endowed with many negative or positive charges on their surface. In this case, the electrostatic repulsive forces dominate, making the nanofibers entirely dispersed [79]. As the colloidal and rheological behaviors of Pal and Sep suspensions are less affected by common pH conditions (pH = 4–11), Pal and Sep are more popular in WBDFs for saline formations or offshore drilling. In practice, the PZC, as well as other properties of different Pal and Sep dispersions, may vary considerably. The response of these fibrous clay minerals to the pH conditions may also be different. Therefore, screening of different mineral resources is required.

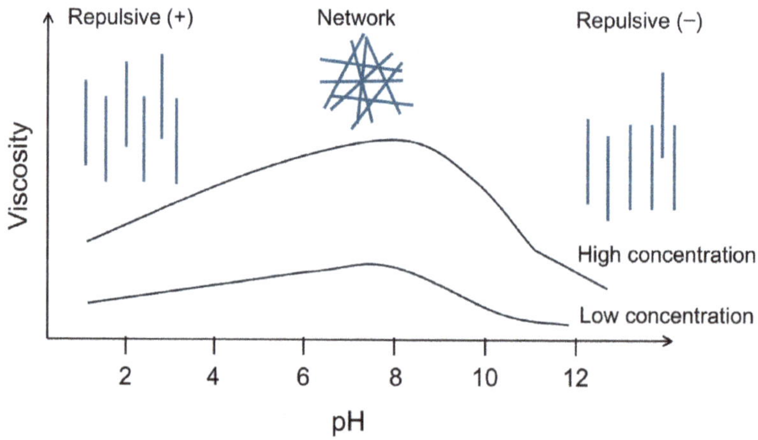

Figure 4. General description of the influence of the pH level on the viscosity of nanofibrous clay minerals.

3.5. Influence of Electrolytes

Although Pal- and Sep-based WBDFs have better salt resistance than those fluids based on Mt, the effect of electrolytes on the rheological properties of Pal and Sep fluids cannot be ignored. Generally, the salt resistance of Pal and Sep is related to the surface charge and pH conditions. The rheological behavior of minerals with smaller charge densities is less affected by the electrolytes, whereas those with larger charge densities exhibit a more pronounced response. Typically, Pal and Sep exhibit the best salt resistance between pH levels of 4 and 8 [53]. In summary, electrolytes influence the rheological behaviors of Pal and Sep dispersions by compressing the electrical double layer, reducing electrostatic repulsion, and enhancing van der Waals attraction, which leads to increased fiber aggregation and network formation [81].

It is worth noting that although Pal and Sep have better salt resistance in WBDFs, most of the previous studies have been conducted using single-electrolyte systems (e.g., NaCl or KCl). There is a lack of comprehensive studies on systems with complex salts (e.g., brine-based drilling fluids and seawater-based drilling fluids). In recent years, brine-based and seawater-based drilling fluids have become increasingly crucial high-performance drilling fluids. Therefore, it is necessary to carry out an in-depth study on the rheological behavior and rheology control mechanism of fibrous clay minerals in complex electrolyte systems.

3.6. *Influence of Temperatures*

The extraction of oil and gas from deep and ultra-deep earth has increased steadily in recent years [82,83]. As the depth of a reservoir location increases, the corresponding pressure and temperature also increase. Reservoirs are classified as normal, high-temperature and high-pressure (HTHP) (150 °C, 68.95 MPa), ultra-HPHT (205 °C, 137.9 MPa), and extremely HPHT (260 °C, 241.32 MPa) reservoirs [84]. Consequently, drilling operations in HTHP fields are inherently risky and challenging. The management of rheological and filtration properties represents a significant factor contributing to the high cost of HTHP drilling operations [85].

Pal and Sep exhibit good thermal stability in WBDFs due to their unique fibrous structures. Studies have shown that Pal and Sep maintain their structural integrity and rheological properties at temperatures up to 200 °C [36,40–42,47,51,59,86–89]. This thermal stability ensures that WBDFs retain their viscosity and gel strength, which are essential for effective cutting transport and wellbore stability. However, the thermal stability of these clay minerals can be affected by the presence of electrolytes and other additives in a drilling fluid. Electrolytes can alter the surface charge of the clay particles, potentially leading to changes in colloidal interactions and aggregation behavior at elevated temperatures [90]. Although Pal and Sep show remarkable rheological properties at temperatures below 200 °C, the thermal stability of Pal and Sep in real WBDFs at higher temperatures (e.g., 250 °C) is unclear. In addition, high-temperature wells are generally accompanied by high pressure. Therefore, it is necessary to consider the effects of both temperature and pressure on the rheological behavior of these two clay minerals, especially at ultra-high temperatures.

4. Filtration Properties

The filtration properties of Pal and Sep in WBDFs are critical for maintaining wellbore stability and minimizing fluid loss into the formation. Studies have shown that Pal and Sep present more considerable fluid loss than Mt in WBDFs [9,38,41,63,86,87,91–94]. For instance, a real WBDF prepared with 4% Pal showed a fluid loss of 26.6 mL after aging at 180 °C for 16 h, whereas a control drilling fluid with 4% commercial bentonite exhibited a fluid loss of 16.4 mL [93]. This is due to the fibrous characteristic of Pal and Sep, which forms a filter cake with many pores that allow water to permeate easily (Figure 5).

In order to improve the filtration properties of Pal and Sep in WBDFs, filtration control additives, such as cellulose, starches, and humic acids, are generally considered to be used [86,90,95]. It has been shown that adding polymeric filter loss reducers improves the filtration loss of Pal and Sep in WBDFs significantly, especially at high temperatures [89,96–98]. The use of polymers in drilling fluids is already widespread, and they are mainly used as rheology modifiers and fluid loss reducers. In controlling filtration loss, polymers generally interact with the surfaces of clay minerals to make the structure of the filter cake more compact, thereby preventing the loss of liquid. However, previous studies have all been about the interaction between Mt and polymers. Pal and Sep are quite different from Mt in terms of structure and properties. There is a need for detailed studies on the interaction of nanofibrous clay minerals with typical filter loss-reducing agents and their influence on the properties of a filter cake and the amount of fluid loss. Developing special filter loss-reducing agents for drilling fluids may be necessary based on Pal and Sep. Furthermore, it was demonstrated that the addition of nano Sep to a conventional WBDF (with Mt as the main rheological additive) could improve the overall stability of the drilling fluid by enhancing the dispersion and suspension of fine particles, thereby forming a low-permeability filter cake to decrease fluid loss.

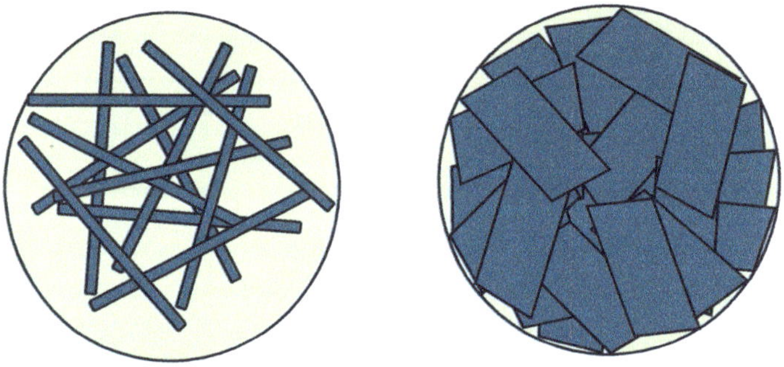

(a) (b)

Figure 5. Diagrams of the filter cake structures of (**a**) nanofibrous clay minerals and (**b**) layered Mt.

5. Method of Applying Pal and Sep in WBDFs

Pal and Sep are used in WBDFs in a similar way to bentonite (i.e., clay minerals are directly mixed with other additives in water). Unlike bentonite, Pal and Sep do not need to be modified with soda or other sodium salts. Despite the remarkable rheological properties and thermal stability in WBDFs, their use as standalone agents is not recommended due to insufficient fluid loss control. Another reason is that Pal and Sep may display worse rheological properties than Mt in an aqueous system [99,100]. Pal and Sep are typically incorporated into WBDFs at a concentration based on Mt, as they have been demonstrated to enhance the salt resistance of drilling fluids. Therefore, in order to balance the rheological properties of Mt with the salt resistance of fibrous clay minerals, a mixture of layered Mt and fibrous clay minerals is generally used. It has been demonstrated that this method can effectively improve the rheological properties and stability of WBDFs under HTHP conditions [36,38,39,93,101,102]. The synergetic use of fibrous and layered clay minerals in WBDFs is typically achieved by combining the two types of clay minerals independently. However, natural Mt may be deposited with some Pal or Sep impurities in some situations. Such blended clays can be directly employed as the rheological additives for WBDFs, improving the rheological properties of drilling fluids [103]. This is important when using mixed clay resources because separation and purification may be unnecessary.

Although the use of layered and nanofibrous clay minerals in combination represents a promising avenue for enhancing the rheological and filtration characteristics of WBDFs, the optimal proportion and ultimate properties are contingent upon the provenance, physical attributes, and processing techniques employed for the clay minerals. For example, Alghareeb [9] found that adding Sep increased the viscosity and yield stress of bentonite dispersion, contrary to the findings of previous investigations [99,100]. Therefore, gaining a deeper understanding of the microstructure and colloidal forces which govern the clay mineral complexes in drilling fluids is essential. Unfortunately, a detailed study of the microstructure of mixed-layer fibrous clay mineral dispersions has yet to be conducted. Recently, the microstructure of mixed Laponite-Sep dispersions was revealed [79], and this may be helpful in understanding the microstructures of Mt-Pal and Mt-Sep mixtures. In that report, nanofibers were observed to be randomly orientated and interact attractively in a cross-fiber configuration, forming a three-dimensional network structure. The predominant interaction configuration appeared to be between the positive edges of Laponite layers (or flaky particles) and the negative silica face of Sep fibers. Mt probably interacts with Pal or Sep following a similar mechanism. However, it should be noticed that Mt particles are much larger than Laponite flakes (Figure 6). The Mt-Pal and Mt-Sep interactions may differ from the Laponite-Sep interaction. Consequently, further investigations are required to elucidate the microstructures and colloidal chemistry of Mt-Pal and Mt-Sep mixtures in WBDFs.

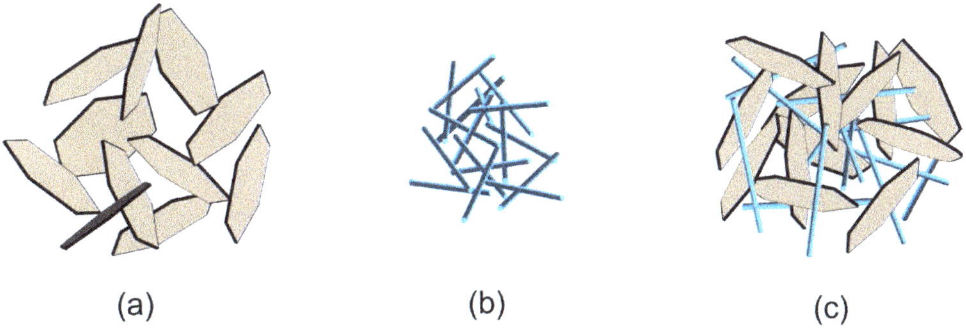

Figure 6. Structures of (**a**) Mt dispersion, (**b**) fibrous clay mineral dispersion, and (**c**) mixed-layer fibrous clay mineral dispersion.

6. Challenges

6.1. Cost

One of the significant challenges in the application of nanofibrous clay minerals in WBDFs is the cost. Although these minerals offer unique properties which can enhance drilling fluid performance, their production and processing costs can be relatively high compared with conventional additives. For example, the prices of Pal and Sep usually range from USD 200 to 600 per ton, while bentonite ranges from USD 50 to 150 per ton. In addition, the refinement and activation processes for optimizing the rheological and filtration properties of nanofibrous clay minerals also increase the total cost. Moreover, scaling up from laboratory to industrial production can introduce further financial barriers. In order to make these nanofibrous minerals more economically viable, it is necessary to develop cost-effective processing techniques and gain a deeper understanding of how to maximize their performance with minimal quantities.

6.2. Demand and Supply

The demand for Pal and Sep in the drilling industry is growing as operators seek more efficient and environmentally friendly drilling fluid additives. However, the supply of these minerals can be inconsistent, depending on geographical availability and mining conditions. Pal and Sep deposits are not uniformly distributed worldwide, leading to potential supply chain issues, especially in regions where these minerals are not readily available. This uneven distribution can cause price volatility and supply bottlenecks, hindering widespread adoption. More importantly, the properties of natural minerals give rise to enormous variations in their genesis, deposits, and chemical compositions. Such variations can further reinforce the imbalance between supply and demand for Pal and Sep.

6.3. Environmental Concerns

The environmental impact of using nanofibrous clay minerals in drilling fluids is a critical consideration. While these minerals are naturally occurring and generally considered environmentally benign, the processes involved in their extraction, refinement, and disposal can raise environmental concerns. For instance, acids or surfactants may be employed to remove impurities (e.g., carbonates and quartz). The mining of Pal and Sep can lead to land degradation and habitat destruction if not managed sustainably. Additionally, the disposal of drilling fluids containing these minerals poses challenges, particularly in terms of potential contamination of soil and water bodies. Research into more sustainable mining practices, along with the development of biodegradable or easily recyclable drilling fluid formulations, is necessary to address these environmental concerns.

6.4. Technical Requirements

From a technical perspective, the integration of nanofibrous clay minerals into water-based drilling fluids requires careful consideration of several factors, including compatibility with other fluid components, optimal concentration levels, and the impact on fluid properties under various drilling conditions (particularly under high temperatures and high pressures). The unique fibrous structure of Pal and Sep can significantly influence the rheological behavior of drilling fluids, but achieving the desired balance of viscosity, yield stress, and filtration control is complex. Moreover, the effectiveness of these minerals can vary depending on the specific conditions of the drilling operation, such as temperature, pressure, and salinity. Addressing these technical challenges demands extensive research and field trials to develop standardized formulations which can consistently deliver the desired performance across different drilling scenarios.

6.5. Field Application

Field application of nanofibrous clay minerals in drilling fluids presents both opportunities and challenges. While laboratory studies have demonstrated their potential benefits, translating these findings into real-world drilling operations requires overcoming practical obstacles. Deployment of these minerals in the field must account for the variability in drilling environments, including different geological formations and operational parameters. Ensuring consistent dispersion and performance in large-scale operations requires the development of specialized equipment and procedures. Furthermore, real-time monitoring and adjustment of drilling fluid properties are essential to optimize the use of Pal and Sep in diverse field conditions. Despite these challenges, successful field applications can lead to significant improvements in drilling efficiency and environmental compliance.

7. Conclusions and Perspectives

Overall, the application of nanofibrous clay minerals, specifically Pal and Sep, in water-based drilling fluids presents significant potential for enhancing the performance and sustainability of drilling operations. These minerals offer unique rheological properties, thermal stability, and filtration control which can improve the efficiency of drilling fluids under a variety of challenging conditions. However, their widespread adoption is currently limited by several factors, including high costs, supply chain variability, environmental concerns, and technical challenges related to their integration into drilling fluid formulations. Despite these obstacles, ongoing research and technological advancements suggest that these challenges can be mitigated with appropriate strategies, paving the way for broader application in the industry.

Regarding weaknesses and challenges, the following suggestions are proposed:

(1) It is recommended that future research focuses on developing more cost-effective processing methods for Pal and Sep. This could include optimizing the refinement processes to reduce energy consumption and exploring the potential of using smaller quantities of these minerals in drilling fluid formulations without compromising performance. Furthermore, developing new processing technologies to enhance low-grade or underperforming minerals is essential to reducing costs and ensuring a sufficient supply.

(2) To mitigate environmental concerns, it is crucial to promote sustainable mining practices which minimize the ecological footprint of extracting Pal and Sep. Additionally, research should explore the development of environmentally friendly drilling

fluid formulations which incorporate these minerals. This could include designing biodegradable or easily recyclable fluids, thereby reducing the environmental impact of their disposal.

(3) In terms of technical requirements, further research is needed to fully understand the interactions between nanofibrous clay minerals and other drilling fluid components. Developing standardized formulations which are optimized for different drilling conditions will be essential for ensuring consistent performance in the field. Collaboration between academia, industry, and field operators could accelerate the development of these standardized formulations and their testing in real-world conditions.

Author Contributions: Conceptualization, G.Z.; writing—original draft preparation, G.Z., J.Z., and J.C.; writing—review and editing, Q.L. (Qian Liu), W.F., and Q.L. (Qiang Li). All authors have read and agreed to the published version of the manuscript.

Funding: This research was funded by the Guangdong Basic and Applied Basic Research Foundation (grant number 2019A1515110263), and the APC was funded by Guangdong University of Technology.

Data Availability Statement: The original contributions presented in this study are included in the article. Further inquiries can be directed to the corresponding author.

Conflicts of Interest: The authors declare no conflicts of interest.

References

1. Li, Q.; Zhuang, G.; Yuan, P.; Bergaya, F. Chapter 12—Future challenges related to clay minerals in drilling and drilling fluids. In *Clay Science in Drilling and Drilling Fluids*; Zhuang, G., Yuan, P., Eds.; Elsevier: Amsterdam, The Netherlands, 2024; pp. 313–338.
2. Zhuang, G.; Li, Q.; Bergaya, F.; Yuan, P. Chapter 1—The significance of clay minerals in drilling and drilling fluids. In *Clay Science in Drilling and Drilling Fluids*; Zhuang, G., Yuan, P., Eds.; Elsevier: Amsterdam, The Netherlands, 2024; pp. 1–19.
3. Caenn, R.; Darley, H.C.H.; Gray, G.R. Chapter 1—Introduction to Drilling Fluids. In *Composition and Properties of Drilling and Completion Fluids*, 7th ed.; Caenn, R., Darley, H.C.H., Gray, G.R., Eds.; Gulf Professional Publishing: Boston, MA, USA, 2017; pp. 1–34.
4. Tabatabaee Moradi, S.S. Development of a water-based drilling fluid for chemical enhancement of drilling rate in a dolomite rock sample. *J. Pet. Sci. Eng.* **2022**, *216*, 110768. [CrossRef]
5. Zhuang, G.; Zhang, Z.; Jaber, M. Organoclays used as colloidal and rheological additives in oil-based drilling fluids: An overview. *Appl. Clay Sci.* **2019**, *177*, 63–81. [CrossRef]
6. Luckham, P.F.; Rossi, S. The colloidal and rheological properties of bentonite suspensions. *Adv. Colloid Interface Sci.* **1999**, *82*, 43–92. [CrossRef]
7. Kok, M.V. Rheological and thermal analysis of bentonites for water base drilling fluids. *Energy Sources* **2004**, *26*, 145–151.
8. Xu, T.; Bezuijen, A. Bentonite slurry infiltration into sand: Filter cake formation under various conditions. *Géotechnique* **2019**, *69*, 1095–1106. [CrossRef]
9. Alghareeb, A. A Review on Controlling Bentonite-Based Drilling Mud Properties. *Int. J. Archit. Energy Urban.* **2020**, *1*, 44–55.
10. Vipulanandan, C.; Mohammed, A. Effect of drilling mud bentonite contents on the fluid loss and filter cake formation on a field clay soil formation compared to the API fluid loss method and characterized using Vipulanandan models. *J. Pet. Sci. Eng.* **2020**, *189*, 107029. [CrossRef]
11. Temraz, M.G.; Hassanien, I. Mineralogy and rheological properties of some Egyptian bentonite for drilling fluids. *J. Nat. Gas Sci. Eng.* **2016**, *31*, 791–799. [CrossRef]
12. Larsen, D.H. Use of Clay in Drilling Fluids. *Clays Clay Miner.* **1952**, *1*, 269–281. [CrossRef]
13. Leong, Y.K.; Du, M.; Au, P.I.; Clode, P.; Liu, J. Microstructure of Sodium Montmorillonite Gels with Long Aging Time Scale. *Langmuir* **2018**, *34*, 9673–9682. [CrossRef]
14. Au, P.-I.; Du, M.; Liu, J.; Haq, M.B.; Leong, Y.-K. Surface chemistry, rheology and microstructure of as-received SHCa-1 hectorite gels. *Clay Miner.* **2019**, *54*, 269–275. [CrossRef]
15. Du, M.; Liu, P.; Clode, P.L.; Liu, J.; Haq, B.; Leong, Y.-K. Impact of additives with opposing effects on the rheological properties of bentonite drilling mud: Flow, ageing, microstructure and preparation method. *J. Pet. Sci. Eng.* **2020**, *192*, 107282. [CrossRef]
16. Du, M.; Liu, P.; Wong, J.-E.; Clode, P.L.; Liu, J.; Leong, Y.-K. Colloidal forces, microstructure and thixotropy of sodium montmorillonite (SWy-2) gels: Roles of electrostatic and van der Waals forces. *Appl. Clay Sci.* **2020**, *195*, 105710. [CrossRef]
17. Caenn, R.; Darley, H.C.H.; Gray, G.R. Chapter 4—Clay Mineralogy and the Colloid Chemistry of Drilling Fluids. In *Composition and Properties of Drilling and Completion Fluids*, 7th ed.; Caenn, R., Darley, H.C.H., Gray, G.R., Eds.; Gulf Professional Publishing: Boston, MA, USA, 2017; pp. 93–134.

18. Kelessidis, V.C.; Christidis, G.; Makri, P.; Hadjistamou, V.; Tsamantaki, C.; Mihalakis, A.; Papanicolaou, C.; Foscolos, A. Gelation of water–bentonite suspensions at high temperatures and rheological control with lignite addition. *Appl. Clay Sci.* **2007**, *36*, 221–231. [CrossRef]
19. Kelessidis, V.C.; Tsamantaki, C.; Dalamarinis, P. Effect of pH and electrolyte on the rheology of aqueous Wyoming bentonite dispersions. *Appl. Clay Sci.* **2007**, *38*, 86–96. [CrossRef]
20. Kelessidis, V.C.; Tsamantaki, C.; Michalakis, A.; Christidis, G.E.; Makri, P.; Papanicolaou, K.; Foscolos, A. Greek lignites as additives for controlling filtration properties of water–bentonite suspensions at high temperatures. *Fuel* **2007**, *86*, 1112–1121. [CrossRef]
21. Kelessidis, V.C.; Maglione, R. Yield stress of water–bentonite dispersions. *Colloids Surf. A Physicochem. Eng. Asp.* **2008**, *318*, 217–226. [CrossRef]
22. Chen, J.S.; Cushman, J.H.; Low, P.F. Rheological Behavior of Na-Montmorillonite Suspensions at Low Electrolyte Concentration. *Clays Clay Miner.* **1990**, *38*, 57–62. [CrossRef]
23. Santoyo, E.; Santoyo-Gutiérrez, S.; García, A.; Espinosa, G.; Moya, S.L. Rheological property measurement of drilling fluids used in geothermal wells. *Appl. Therm. Eng.* **2001**, *21*, 283–302. [CrossRef]
24. Liu, J.; Cheng, Y.; Zhou, F.; Amutenya Evelina, L.M.; Long, W.; Chen, S.; He, L.; Yi, X.; Yang, X. Evaluation method of thermal stability of bentonite for water-based drilling fluids. *J. Pet. Sci. Eng.* **2022**, *208*, 109239. [CrossRef]
25. Liu, J.; Zhang, T.; Sun, Y.; Lin, D.; Feng, X.; Wang, F. Insights into the high temperature-induced failure mechanism of bentonite in drilling fluid. *Chem. Eng. J.* **2022**, *445*, 136680. [CrossRef]
26. Afolabi, R.O.; Orodu, O.D.; Seteyeobot, I. Predictive modelling of the impact of silica nanoparticles on fluid loss of water based drilling mud. *Appl. Clay Sci.* **2018**, *151*, 37–45. [CrossRef]
27. Cheraghian, G.; Wu, Q.; Mostofi, M.; Li, M.-C.; Afrand, M.; Sangwai, J.S. Effect of a novel clay/silica nanocomposite on water-based drilling fluids: Improvements in rheological and filtration properties. *Colloids Surf. A Physicochem. Eng. Asp.* **2018**, *555*, 339–350. [CrossRef]
28. Elkatatny, S.; Kamal, M.S.; Alakbari, F.; Mahmoud, M. Optimizing the rheological properties of water-based drilling fluid using clays and nanoparticles for drilling horizontal and multi-lateral wells. *Appl. Rheol.* **2018**, *28*, 43606.
29. Aftab, A.; Ismail, A.R.; Ibupoto, Z.; Akeiber, H.; Malghani, M. Nanoparticles based drilling muds a solution to drill elevated temperature wells: A review. *Renew. Sustain. Energy Rev.* **2017**, *76*, 1301–1313. [CrossRef]
30. Vryzas, Z.; Kelessidis, V.C. Nano-Based Drilling Fluids: A Review. *Energies* **2017**, *10*, 540. [CrossRef]
31. Ali, M.; Jarni, H.H.; Aftab, A.; Ismail, A.R.; Saady, N.M.C.; Sahito, M.F.; Keshavarz, A.; Iglauer, S.; Sarmadivaleh, M. Nanomaterial-Based Drilling Fluids for Exploitation of Unconventional Reservoirs: A Review. *Energies* **2020**, *13*, 3417. [CrossRef]
32. Kosynkin, D.V.; Ceriotti, G.; Wilson, K.C.; Lomeda, J.R.; Scorsone, J.T.; Patel, A.D.; Friedheim, J.E.; Tour, J.M. Graphene oxide as a high-performance fluid-loss-control additive in water-based drilling fluids. *ACS Appl. Mater. Interfaces* **2012**, *4*, 222–227. [CrossRef] [PubMed]
33. Ridha, S.; Ibrahim, A.; Shahari, R.; Fonna, S. Graphene nanoplatelets as high-performance filtration control material in water-based drilling fluids. *IOP Conf. Ser.-Mater. Sci. Eng.* **2018**, *352*, 012025. [CrossRef]
34. Sadeghalvaad, M.; Sabbaghi, S. The effect of the TiO_2/polyacrylamide nanocomposite on water-based drilling fluid properties. *Powder Technol.* **2015**, *272*, 113–119. [CrossRef]
35. Aftab, A.; Ali, M.; Arif, M.; Panhwar, S.; Saady, N.M.C.; Al-Khdheeawi, E.A.; Mahmoud, O.; Ismail, A.R.; Keshavarz, A.; Iglauer, S. Influence of tailor-made TiO_2/API bentonite nanocomposite on drilling mud performance: Towards enhanced drilling operations. *Appl. Clay Sci.* **2020**, *199*, 105862. [CrossRef]
36. Abdo, J.; Haneef, M.D. Nanomaterials Modified Drilling Fluid for Improving Deep Drilling Conditions. *J. Energy Resour. Technol.* **2022**, *144*, 073202. [CrossRef]
37. Abdo, J. Nano-attapulgite for improved tribological properties of drilling fluids. *Surf. Interface Anal.* **2014**, *46*, 882–887. [CrossRef]
38. Abdo, J.; AL-Sharji, H.; Hassan, E. Effects of nano-sepiolite on rheological properties and filtration loss of water-based drilling fluids. *Surf. Interface Anal.* **2016**, *48*, 522–526. [CrossRef]
39. Al-Malki, N.; Pourafshary, P.; Al-Hadrami, H.; Abdo, J. Controlling bentonite-based drilling mud properties using sepiolite nanoparticles. *Pet. Explor. Dev.* **2016**, *43*, 717–723. [CrossRef]
40. Ettehadi, A. A Comparative Study on Thixotropic Behavior of Clay Based Drilling Fluids. *Annu. Trans. Nord. Rheol. Soc.* **2020**, *28*, 99–107.
41. Ettehadi, A. An Experimental Study on Structural and Thermal Stability of Water-Based Drilling Fluids. *Eur. J. Sci. Technol.* **2021**, *23*, 70–80.
42. Ettehadi, A.; Ülker, C.; Altun, G. Nonlinear viscoelastic rheological behavior of bentonite and sepiolite drilling fluids under large amplitude oscillatory shear. *J. Pet. Sci. Eng.* **2022**, *208*, 109210. [CrossRef]
43. Weng, J.; Gong, Z.; Liao, L.; Lv, G.; Tan, J. Comparison of organo-sepiolite modified by different surfactants and their rheological behavior in oil-based drilling fluids. *Appl. Clay Sci.* **2018**, *159*, 94–101. [CrossRef]
44. Buriti, B.; Barsosa, M.E.; Buriti, J.D.; Cartaxo, J.D.; Ferreira, H.S.; Neves, G.D. Modification of palygorskite with cationic and nonionic surfactants for use in oil-based drilling fluids. *J. Therm. Anal. Calorim.* **2022**, *147*, 2935–2945. [CrossRef]

45. Huang, X.; Shen, H.; Sun, J.; Lv, K.; Liu, J.; Dong, X.; Luo, S. Nanoscale Laponite as a Potential Shale Inhibitor in Water-Based Drilling Fluid for Stabilization of Wellbore Stability and Mechanism Study. *ACS Appl. Mater. Interfaces* **2018**, *10*, 33252–33259. [CrossRef] [PubMed]
46. Huang, X.-B.; Sun, J.-S.; Huang, Y.; Yan, B.-C.; Dong, X.-D.; Liu, F.; Wang, R. Laponite: A promising nanomaterial to formulate high-performance water-based drilling fluids. *Pet. Sci.* **2021**, *18*, 579–590. [CrossRef]
47. Galan, E. Properties and applications of palygorskite-sepiolite clays. *Clay Miner.* **1996**, *31*, 443–453. [CrossRef]
48. Zhuang, G.; Zhang, Z.; Bergaya, F.; Yuan, P. Chapter 3—Application of fibrous clay minerals in water-based drilling fluids. In *Clay Science in Drilling and Drilling Fluids*; Zhuang, G., Yuan, P., Eds.; Elsevier: Amsterdam, The Netherlands, 2024; pp. 55–82.
49. Neaman, A.; Singer, A. Possible use of the Sacalum (Yucatan) palygorskite as drilling muds. *Appl. Clay Sci.* **2004**, *25*, 121–124. [CrossRef]
50. Neaman, A.; Singer, A. Rheological properties of aqueous suspensions of palygorskite. *Soil Sci. Soc. Am. J.* **2000**, *64*, 427–436. [CrossRef]
51. Echt, T.; Plank, J. An improved test protocol for high temperature carrying capacity of drilling fluids exemplified on a sepiolite mud. *J. Nat. Gas Sci. Eng.* **2019**, *70*, 102964. [CrossRef]
52. García-Villén, F.; Sánchez-Espejo, R.; López-Galindo, A.; Cerezo, P.; Viseras, C. Design and characterization of spring water hydrogels with natural inorganic excipients. *Appl. Clay Sci.* **2020**, *197*, 105772. [CrossRef]
53. Liu, P.; Du, M.; Clode, P.; Li, H.; Liu, J.; Leong, Y.K. Surface Chemisty, Microstructure, and Rheology of Thixotropic 1-D Sepiolite Gels. *Clays Clay Miner.* **2020**, *68*, 9–22. [CrossRef]
54. Choupani, M.A.; Tabatabaee Moradi, S.S.; Tabatabaei Nejad, S.A. Study on Attapulgite as Drilling Fluid Clay Additive in Persian Gulf Seawater. *Int. J. Eng.* **2022**, *35*, 587–595.
55. Guggenheim, S.; Krekeler, M.P.S. Chapter 1—The Structures and Microtextures of the Palygorskite–Sepiolite Group Minerals. In *Developments in Clay Science*; Galàn, E., Singer, A., Eds.; Elsevier: Amsterdam, The Netherlands, 2011; pp. 3–32.
56. Álvarez, A.; Santarén, J.; Esteban-Cubillo, A.; Aparicio, P. Chapter 12—Current Industrial Applications of Palygorskite and Sepiolite. In *Developments in Clay Science*; Galàn, E., Singer, A., Eds.; Elsevier: Amsterdam, The Netherlands, 2011; pp. 281–298.
57. García-Romero, E.; Suárez, M. Sepiolite–palygorskite: Textural study and genetic considerations. *Appl. Clay Sci.* **2013**, *86*, 129–144. [CrossRef]
58. Murray, H.H. Applied clay mineralogy today and tomorrow. *Clay Miner.* **1999**, *34*, 39–49. [CrossRef]
59. Murray, H.H. Traditional and new applications for kaolin, smectite, and palygorskite: A general overview. *Appl. Clay Sci.* **2000**, *17*, 207–221. [CrossRef]
60. Haden, W. Attapulgite: Properties and Uses. *Clays Clay Miner.* **1961**, *10*, 284–290. [CrossRef]
61. Haden, W.L.; Schwint, I.A. Attapulglte: Its Properties and Applications. *Ind. Eng. Chem.* **1967**, *59*, 58–69. [CrossRef]
62. Cui, J.; Zhang, Z.; Han, F. Effects of pH on the gel properties of montmorillonite, palygorskite and montmorillonite-palygorskite composite clay. *Appl. Clay Sci.* **2020**, *190*, 105543. [CrossRef]
63. Santanna, V.C.; Silva, S.L.; Silva, R.P.; Castro Dantas, T.N. Use of palygorskite as a viscosity enhancer in salted water-based muds: Effect of concentration of palygorskite and salt. *Clay Miner.* **2020**, *55*, 48–52. [CrossRef]
64. Dahab, A.S. Thermal-Stability of Drilling-Fluids Prepared from Saudi Palygorskite. *J. Can. Pet. Technol.* **1991**, *30*, 49–52. [CrossRef]
65. Baltar, C.A.M.; da Luz, A.B.; Baltar, L.M.; de Oliveira, C.H.; Bezerra, F.J. Influence of morphology and surface charge on the suitability of palygorskite as drilling fluid. *Appl. Clay Sci.* **2009**, *42*, 597–600. [CrossRef]
66. Xu, J.; Zhang, J.; Wang, Q.; Wang, A. Disaggregation of palygorskite crystal bundles via high-pressure homogenization. *Appl. Clay Sci.* **2011**, *54*, 118–123. [CrossRef]
67. Xu, J.; Wang, W.; Wang, A. Superior dispersion properties of palygorskite in dimethyl sulfoxide via high-pressure homogenization process. *Appl. Clay Sci.* **2013**, *86*, 174–178. [CrossRef]
68. Xu, J.; Wang, W.; Wang, A. Effects of solvent treatment and high-pressure homogenization process on dispersion properties of palygorskite. *Powder Technol.* **2013**, *235*, 652–660. [CrossRef]
69. Xu, J.X.; Wang, W.B.; Wang, A.Q. Dispersion of palygorskite in ethanol-water mixtures via high-pressure homogenization: Microstructure and colloidal properties. *Powder Technol.* **2014**, *261*, 98–104. [CrossRef]
70. Boudriche, L.; Chamayou, A.; Calvet, R.; Hamdi, B.; Balard, H. Influence of different dry milling processes on the properties of an attapulgite clay, contribution of inverse gas chromatography. *Powder Technol.* **2014**, *254*, 352–363. [CrossRef]
71. Viseras, C.; Meeten, G.H.; Lopez-Galindo, A. Pharmaceutical grade phyllosilicate dispersions: The influence of shear history on floc structure. *Int. J. Pharm.* **1999**, *182*, 7–20. [CrossRef] [PubMed]
72. Darvishi, Z.; Morsali, A. Sonochemical preparation of palygorskite nanoparticles. *Appl. Clay Sci.* **2011**, *51*, 51–53. [CrossRef]
73. Pardo-Canales, L.; Essih, S.; Cecilia, J.A.; Domínguez-Maqueda, M.; Olmo-Sánchez, M.I.; Pozo-Rodríguez, M.; Franco, F. Modification of the textural properties of palygorskite through microwave assisted acid treatment. Influence of the octahedral sheet composition. *Appl. Clay Sci.* **2020**, *196*, 105745. [CrossRef]
74. Myriam, M.; Suárez, M.; Martín-Pozas, J.M. Structural and Textural Modifications of Palygorskite and Sepiolite Under Acid Treatment. *Clays Clay Miner.* **1998**, *46*, 225–231. [CrossRef]
75. Oliveira, R.; Acchar, W.; Soares, G.; Barreto, L. The increase of surface area of a Brazilian palygorskite clay activated with sulfuric acid solutions using a factorial design. *Mater. Res.* **2013**, *16*, 924–928. [CrossRef]

76. Zhou, H.; Murray, H.H. Chapter 10—Overview of Chinese Palygorskite Clay Resources—Their Geology, Mineralogy, Depositional Environment, Applications and Processing. In *Developments in Clay Science*; Galàn, E., Singer, A., Eds.; Elsevier: Amsterdam, The Netherlands, 2011; pp. 239–263.
77. Altun, G.; Serpen, U. Investigating improved rheological and fluid loss performance of sepiolite muds under elevated temperatures. In Proceedings of the World Geothermal Congress, Antalya, Turkey, 24–29 April 2005.
78. Cinar, M.; Can, M.F.; Sabah, E.; Karaguzel, C.; Celik, M.S. Rheological properties of sepiolite ground in acid and alkaline media. *Appl. Clay Sci.* **2009**, *42*, 422–426. [CrossRef]
79. Liu, P.; Du, M.; Clode, P.; Liu, J.; Leong, Y.-K. Rod–plate interactions in sepiolite–LAPONITE® gels: Microstructure, surface chemistry and rheology. *Soft Matter* **2021**, *17*, 2614–2623. [CrossRef] [PubMed]
80. Alves, L.; Ferraz, E.; Santarén, J.; Rasteiro, M.G.; Gamelas, J.A.F. Improving Colloidal Stability of Sepiolite Suspensions: Effect of the Mechanical Disperser and Chemical Dispersant. *Minerals* **2020**, *10*, 779. [CrossRef]
81. Sabah, E.; Mart, U.; Çınar, M.; Çelik, M.S. Zeta Potentials of Sepiolite Suspensions in Concentrated Monovalent Electrolytes. *Sep. Sci. Technol.* **2007**, *42*, 2275–2288. [CrossRef]
82. Gautam, S.; Guria, C.; Rajak, V.K. A state of the art review on the performance of high-pressure and high-temperature drilling fluids: Towards understanding the structure-property relationship of drilling fluid additives. *J. Pet. Sci. Eng.* **2022**, *213*, 110318. [CrossRef]
83. Mao, H.; Yang, Y.; Zhang, H.; Zhang, J.; Huang, Y. A critical review of the possible effects of physical and chemical properties of subcritical water on the performance of water-based drilling fluids designed for ultra-high temperature and ultra-high pressure drilling applications. *J. Pet. Sci. Eng.* **2020**, *187*, 106795. [CrossRef]
84. Belani, A.; Orr, S. A Systematic Approach to Hostile Environments. *J. Pet. Technol.* **2008**, *60*, 34–39. [CrossRef]
85. Mohamed, A.; Salehi, S.; Ahmed, R. Significance and complications of drilling fluid rheology in geothermal drilling: A review. *Geothermics* **2021**, *93*, 102066. [CrossRef]
86. Carney, L.L.; Meyer, R.L. A New Approach to High Temperature Drilling Fields. In Proceedings of the SPE Annual Fall Technical Conference and Exhibition, New Orleans, LA, USA, 3–6 October 1976. SPE-6025-MS.
87. Ettehadi, A.; Altun, G. Extending thermal stability of calcium carbonate pills using sepiolite drilling fluid. *Pet. Explor. Dev.* **2017**, *44*, 477–486. [CrossRef]
88. Ettehadi, A.; Altun, G. In-Situ Thermal Rheological Properties of Drilling Muds. In Proceedings of the SPE/IADC Middle East Drilling Technology Conference and Exhibition, Abu Dhabi, United Arab Emirates, 29–31 January 2018. SPE/IADC-189349-MS.
89. Altun, G.; Osgouei, A.E.; Ozyurtkan, M.H. Customization of sepiolite based drilling fluids at high temperatures. In Proceedings of the 48th US Rock Mechanics/Geomechanics Symposium, Minneapolis, MN, USA, 1–4 June 2014. ARMA 14-7031: American Rock Mechanics Association.
90. Carney, L.L.; Guven, N. Investigation of changes in the structure of clays during hydrothermal study of drilling fluids. *Soc. Pet. Eng. J.* **1980**, *20*, 385–390. [CrossRef]
91. Zhao, X.; Li, W.; Li, S.; Ji, Y. Application of Several Kinds of Clays and a New Type of Nano-Modified Bontonite in Drilling Fluids. *Adv. Mater. Res.* **2013**, *746*, 489–495. [CrossRef]
92. Tao, Z.; Tianqi, L.; Yuechao, L.; Yang, L.; Gaowei, G.; Jingcheng, C.; Fengshan, Z. Preparation and Rheological Properties of Attapulgite Gel for Aqueous Suspensions. In Proceedings of the 2016 7th International Conference on Education, Management, Computer and Medicine (EMCM 2016), Shenyang, China, 29–31 December 2016.
93. Wenlong, Z.; Xiaoming, W.; Yuming, H.; Jie, X.; Wenshi, W. Research and Application of High-Temperature Drilling Fluid for Scientific Core Drilling Project. In Proceedings of the Abu Dhabi International Petroleum Exhibition & Conference, Abu Dhabi, United Arab Emirates, 13 November 2017. SPE-188906-MS.
94. Christidis, G.E.; Athanasakis, N.; Marinakis, D. Rheological properties of magnesium bentonite and sepiolite suspensions after dynamic ageing at high temperatures. *Clay Miner.* **2024**. [CrossRef]
95. Guven, N.; Panfil, D.J.; Carney, L.L. Comparative Rheology of Water-Based Drilling Fluids with Various Clays. In Proceedings of the International Meeting on Petroleum Engineering, Tianjin, China, 31 October 1988. SPE-17571-MS.
96. Osgouei, A.E.; Ozyurtkan, M.H.; Altun, G.; Dilsiz, E.A. Dynamic Filtration Properties of Clay Based Drilling Muds under Elevated Temperatures. In Proceedings of the SPE Kuwait International Petroleum Conference and Exhibition, Kuwait City, Kuwait, 10–12 December 2012. SPE-163325-MS.
97. Osgouei, A.E.; Ozyurtkan, M.H.; Altun, G. Dynamic Filtration Properties of Fresh Water Sepiolite-based Muds. *Energy Sources Part A Recovery Util. Environ. Eff.* **2014**, *36*, 2079–2086. [CrossRef]
98. Altun, G.; Osgouei, A.E. Investigation and remediation of active-clay contaminated sepiolite drilling muds. *Appl. Clay Sci.* **2014**, *102*, 238–245. [CrossRef]
99. Neaman, A.; Singer, A. Rheology of mixed palygorskite-montmorillonite suspensions. *Clays Clay Miner.* **2000**, *48*, 713–715. [CrossRef]
100. İşçi, E.; Turutoğlu, S.İ. Stabilization of the mixture of bentonite and sepiolite as a water based drilling fluid. *J. Pet. Sci. Eng.* **2011**, *76*, 1–5. [CrossRef]
101. Abdo, J.; Haneef, M.D. Nano-Enhanced Drilling Fluids: Pioneering Approach to Overcome Uncompromising Drilling Problems. *J. Energy Resour. Technol.* **2012**, *134*, 014501. [CrossRef]

102. Abdo, J.; Haneef, M.D. Clay nanoparticles modified drilling fluids for drilling of deep hydrocarbon wells. *Appl. Clay Sci.* **2013**, *86*, 76–82. [CrossRef]
103. Huang, W.; Leong, Y.-K.; Chen, T.; Au, P.-I.; Liu, X.; Qiu, Z. Surface chemistry and rheological properties of API bentonite drilling fluid: pH effect, yield stress, zeta potential and ageing behaviour. *J. Pet. Sci. Eng.* **2016**, *146*, 561–569. [CrossRef]

minerals

MDPI

Article

Application of Hydrothermal Carbon/Bentonite Composites in Improving the Thermal Stability, Filtration, and Lubrication of Water-Based Drilling Fluids

Yubin Zhang [1,2], Daqi Li [1,2], Xianguang Wang [1,2], Changzhi Chen [3,4] and Hanyi Zhong [3,4,*]

1 Sinopec Key Laboratory of Ultra-Deep Well Drilling Engineering Technology, Beijing 102206, China; 15853291860@163.com (Y.Z.); lidq.sripe@sinopec.com (D.L.); wangxg.sripe@sinopec.com (X.W.)
2 SINOPEC Research Institute of Petroleum Engineering, Co., Ltd., Beijing 102206, China
3 State Key Laboratory of Deep Oil and Gas, China University of Petroleum (East China), Qingdao 266580, China; 18753125538@163.com
4 School of Petroleum Engineering, China University of Petroleum (East China), Qingdao 266580, China
* Correspondence: zhonghanyi@126.com; Tel.: +86-18853248274

Abstract

With the increasing harsh drilling environments encountered more frequently than ever before, developing environmentally benign and multifunctional additives is essential to formulate high performance drilling fluids. Herein, hydrothermal carbon/bentonite composites (HCBCs) were prepared by a hydrothermal carbonization reaction using soluble starch and sodium bentonite as raw materials. A systematic investigation was conducted into the effects of HCBC concentration on the rheological, filtration, and lubricating characteristics of xanthan gum, modified starch, and high-temperature polymer slurries. These properties were evaluated before and after exposure to hot rolling at different temperatures. The hydroxyl radical scavenging properties of HCBC were evaluated. Observation showed plentiful micro- and nano-sized carbon spheres deposited on the bentonite particles, endowing the bentonite with better dispersion. HCBCs could maintain stability of the water-based drilling fluids' rheological profile, decrease filtration loss, and improve the lubrication with relatively low concentrations. The excellent properties were attributed to the highly efficient scavenging of free radicals and the stabilization of bentonite particle dispersion.

Keywords: water-based drilling fluid; high temperature degradation; hydrothermal carbon/bentonite composites; radical scavenging; properties stability

1. Introduction

With the increasing demand on resources, complex formations such as deep and ultradeep reservoirs, unconventional shale oil and shale gas, and geothermal reservoirs are more frequently encountered than ever before. To address these needs, a new generation of high-performance water-based drilling fluid is required—one that delivers better hole cleaning, enhanced wellbore stability, stable rheology, and low filtration under high-temperature conditions, reflecting the performance attributes and benefits of oil-based drilling fluids [1,2].

Water-based drilling fluid typically contains bentonite and polymers to impart the desired rheological and filtration properties [3]. The main component of bentonite is montmorillonite. Montmorillonite has a 2:1 layered crystal structure, composed of an aluminum–oxygen octahedral sheet sandwiched between two silicon–oxygen tetrahedral sheets, and contains exchangeable cations (such as Na^+, K^+, Ca^{2+}, etc.). Many unit cells are stacked over each other and form layered structures [4]. Due to the isomorphic replacement of cations within the structure, the crystal surface of montmorillonite is negatively charged. Additionally, the fracturing of Al-O and Si-O result in the electrical properties of end surfaces [5]. When dispersed in an aqueous environment, house of card structures are formed by montmorillonite particles via edge-face interactions. The colloidal behavior of bentonite clay is governed by several factors including surface charge nature, cation exchange capacity (CEC), concentration, particle size and shape, and density [6]. The dispersion of clay particles is a primary cause of water-based drilling fluid instability under high-temperature conditions [7]. A previous study indicated that bentonite tends to flocculate at temperatures above 121 °C [8].

The colloidal stability of a bentonite dispersion can be improved by modifying its surface charge or by providing a steric barrier [3]. Various polymers have been employed during the past decades for different purposes. By adsorbing onto bentonite surfaces, polymers confer effective steric stabilization, preventing aggregation of the dispersed particles. Therefore, the rheology and filtration properties are improved. In addition, polymers can be used as lubricants, emulsifiers, and drag reducers, and are multi-functional [9]. In general, there are natural polymer derivatives and synthetic polymers. When the temperature exceeds 150 °C, biopolymers undergo degradation through mechanisms including thermo-oxidative degradation, hydrolysis, and fragmentation, resulting in complete loss of function [10]. In terms of synthetic polymers, the carbon–carbon backbones provide improved thermal stability; however, thermal degradation still occurs due to oxidation and side-chain hydrolysis [1]. Once the polymers degrade, the colloidal stability of the bentonite dispersion would rapidly be lost. Therefore, developing new additives to improve the colloidal stability of bentonite dispersion under elevated temperatures is important and also a challenging task.

Apart from designing high-temperature-resistant polymers, such as star polymers, comb polymers, and dendritic polymers, combining the advantages of inorganic materials and organic materials to form a composite is also a desirable method. For example, an organic–inorganic composite for high-temperature-, high-salt-resistant water-based drilling fluids was prepared by copolymerizing acrylamide (AM), 2-acrylamido-2-methylpropanesulfonic acid (AMPS), diallydimethylammonium chloride (DMDAAC), and 4-acryloylmorpholine (ACMO) with laponite via aqueous radical polymerization. The water-based drilling fluid containing 2 wt% composite maintained excellent rheology and filtration properties, even after hot rolling at 180 °C and contamination with 15 wt% NaCl [11]. In another study, Ahmed et al. prepared a $SiO_2/g\text{-}C_3N_4$ hybrid and revealed that the hybrid nanoparticles can improve the thermal stability of drilling fluid and reduce the filtration both before and after hot rolling at 225 °F [12]. A zwitterionic silica-based hybrid nanomaterial (ZSHNM) was designed with a dual-functional structure: a silicate core for thermal stability and a zwitterionic shell to improve filtration by complexing with cations and suppressing bentonite aggregation. The addition of 2 wt% ZSHNM enabled the water-based drilling fluid to maintain extraordinary filtration loss properties after being thermally aged at 240 °C [13].

Due to exceptional chemical, physical, mechanical, and thermal attributes, carbon nanomaterials have been paid increasing attention to drastically improve the rheology, filtration, lubrication, and inhibition of drilling fluids [12,14–16]. When bentonite and biomass are mixed together and exposed to a subcritical water, hydrothermal carbonization occurs, and a composite materials of hydrothermal carbon spheres supported on montmorillonite can be obtained. Many oxygen-containing groups including hydroxyl, carboxyl, and carbonyl groups are anchored on the surface of the composites, which improves the dispersion stability of bentonite. Due to these special properties, the composites have been widely used for adsorption [17–19].

Herein, hydrothermal carbon bentonite composites (HCBCs) were designed and fabricated using the low-cost materials of soluble starch and sodium bentonite as the raw materials via one-step hydrothermal carbonization. Fourier transform infrared spectroscopy (FTIR) analysis, thermogravimetric analysis (TGA), X-ray photoelectron spectroscopy (XPS), scanning electron microscope (SEM), transmission electron microscope (TEM), and specific surface area were used to characterize the structure characteristics of HCBCs. A systematic investigation was then conducted to assess the impact of HCBCs on the water-based drilling fluid's key properties, including rheology, filtration, and lubrication.

2. Materials and Methods

2.1. Materials

Sodium-based bentonite was purchased from Shengli Oilfield Boyou Mud Technology Co., Ltd., (Dongying, China), containing 60% montmorillonite, 32% quartz, and 8% potassium feldspar. Methyl violet (MV) (CAS: 8004-87-3), hydrochloric acid (CAS: 7647-01-0), soluble starch (CAS: 9005-84-9), and anhydrous ethanol (CAS: 64-17-5) are analytical grade reagents purchased from China National Pharmaceutical Group Chemical Reagent Co., Ltd., (Beijing, China). Xanthan gum (XC) was purchased from Ordos Zhongxuan Biochemical Co., Ltd., (Zibo, China). Modified starch was provided by China Oilfield Services Limited, (Beijing, China). HT, a high-temperature polymer fluid loss reducer, was provided by Chevron Philips Chemical Company, (The Woodlands, TX, USA).

2.2. Preparation and Characterization of HCBCs

The HCBC was synthesized via a hydrothermal method. Briefly, 4 g of soluble starch and 4 g of sodium bentonite were added to 80 mL of deionized water. The mixture was stirred at 10,000 rpm for 30 min and then subjected to a hydrothermal reaction in a 100 mL Teflon-lined kettle at 200 °C for 16 h. After cooling to room temperature, the product was collected by centrifugation, alternately washed three times with deionized water and absolute ethanol, dried at 80 °C for 24 h, and finally ground and passed through a 100-mesh sieve (Figure 1).

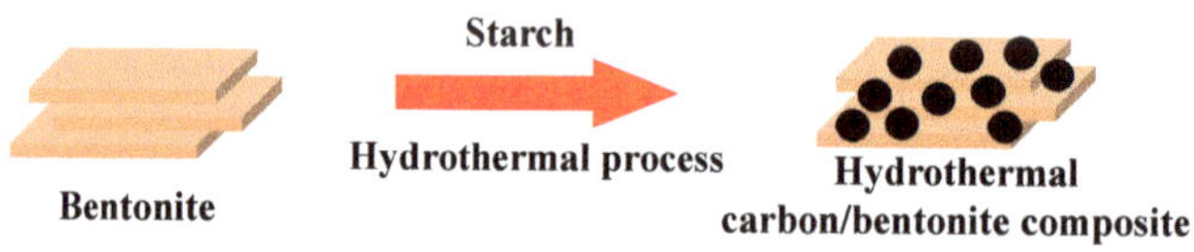

Figure 1. Scheme of hydrothermal reaction.

The functional groups on the surface of the HCBC were investigated with a NEXUS Fourier transform infrared spectrometer, (Thermo Nicolet Corporation, Madison, WI, USA). Thermal stability was evaluated by employing a TGA5500 thermogravimetric analyzer, scanning from 35 to 1000 °C at 10 °C/min under a nitrogen atmosphere. The interlayer spacing of the HCBC sample was measured by X-ray diffraction (XRD) using an X-ray diffractometer (X'Pert PRO MPD), (PAnalytical, Almelo, Overijssel, The Netherlands). The HCBC sample was prepared as a randomly oriented specimen. The measurement was conducted under the following operating conditions: 45 kV (accelerating voltage), 40 mA (tube current), a fixed slit width of 0.76 mm, Cu-Kα radiation (wavelength $\lambda = 0.154$ nm), a scanning rate of 3.86°/min, a step size of 0.017° (2θ), and a scanning range of 3–15°. The microstructure of the HCBC was observed by field emission scanning electron microscopy

(SU8010) (Hitachi, Tokyo, Japan) and high-resolution transmission electron microscopy (FEI Tecnai G2 F20), (FEI NanoPorts, Milwaukee, WI, USA). The specific surface area was tested using an automatic specific surface area analyzer (Micromeritics APSP2460), (Mcmurdik (Shanghai) Instrument Co., Ltd., Shanghai, China). The surface elements of the HCBC were analyzed with a high-resolution X-ray photoelectron spectrometer (ESCALAB XI+), (Thermo Fisher Scientific, Waltham, MA, USA), using Al Kα radiation (hm = 1486.6 eV) during the test, and the binding energy was calibrated by the C1s line at 284.8 eV. The Zeta potential of the HCBC suspension was measured using a multi-angle particle size and high-sensitivity Zeta potential analyzer (Omni), (Brookhaven, New York City, NY, USA). To determine the water adsorption capacity of the HCBC, 1 g HCBC was put in a glass dish. Then the glass dish was placed in the middle position of a dryer where the bottom was filled with water. After sealing the dryer for a certain interval, the glass dish was quickly taken out and weighed. Then, the glass dish with the HCBC was put in the dryer and sealed again until the weight change, reaching a balance.

2.3. *Drilling Fluid Preparation and Properties Measurement*

2.3.1. Drilling Fluid Preparation

Sodium bentonite (16 g) was putted in 400 mL fresh water and mixed at 10,000 rpm for 20 min on a Hamilton Beach mixer. Then, the suspension was sealed and incubated overnight to hydrate and form a bentonite suspension. XC slurry was prepared by adding 0.8 g NaOH and 0.8 g XC into the 400 mL bentonite slurry and stirring at 10,000 rpm for 20 min. The modified starch slurry and HT polymer slurry was prepared via a similar method, where the dosage of both the modified starch and the HT polymer is 4 g.

2.3.2. Drilling Fluid Hot Rolling

The prepared drilling fluid (400 mL) was placed into a stainless-steel aging cell (total volume of 500 mL) and sealed. The aging cell can withstand a maximum temperature of 300 °C and a pressure of 5 MPa (Hengtai Da Electromechanical Equipment Co., Ltd., Qingdao, China). Then the aging cell was transferred into a GW300-X high-temperature roller oven (Xusheng Petroleum Instruments Co., Ltd., Qingdao, China). The drilling fluid was hot-rolled at a specified temperature for 16 h with a rotation speed of 50 r/min.

2.3.3. Drilling Fluid Rheological Properties Measurement

The rheological parameters of the drilling fluid, including plastic viscosity (PV), yield point (YP), and apparent viscosity (AV), were measured using a ZNN-D6 rotational viscometer (Haitongda Special Instrument Co., Ltd., Qingdao, China). The rotational viscometer consists of an inner cylinder (diameter: 34.49 mm) and an outer cylinder (inner diameter: 36.83 mm), with an annular gap of 1.16 mm. It operates at six rotational speeds (3, 6, 100, 200, 300, and 600 r/min), corresponding to six shear rates (5, 10, 170, 340, 511, and 1022 s^{-1}). The experimental procedures strictly adhered to the API standard method API RP 13B-1 (2019 edition) [20]. The sample was poured into the viscometer cup. Measurements were performed sequentially at rotational speeds of 600, 300, 200, 100, 6, and 3 r/min at room temperature. When the dial stabilized, the dial readings were recorded, and the rheological parameters were calculated using the following formulas:

$$AV = \theta_{600}/2 \ (mPa{\cdot}s) \tag{1}$$

$$PV = \theta_{600} - \theta_{300} \ (mPa{\cdot}s) \tag{2}$$

$$YP = AV - PV \ (Pa) \tag{3}$$

For the gel strength measurement, the drilling fluid sample was first stirred at 600 rpm for 10 s, followed by standing for 10 s. Subsequently, the maximum dial reading at a rotation speed of 3 rpm was recorded. Half of this value was recorded as the initial gel strength (gel 10 s). The drilling fluid sample was stirred at 600 rpm for 10 s again and allowed to rest for 10 min. Then, the maximum dial reading at 3 rpm was recorded. Half of this value was recorded as the final gel strength (gel 10 min).

2.4. Hydroxyl Free Radicals Scavenging Experiment

Under the test conditions, MV exhibits its characteristic purple color. The highly reactive hydroxyl radicals generated via the Fenton reaction cause the discoloration of methyl violet, leading to a significant decrease in the maximum absorbance value. With the increase in the dosage of free radical scavenger, the color of the methyl violet solution gradually deepens, and the absorbance increases correspondingly. Therefore, the free radical scavenging efficiency can be indirectly detected [21,22]. First, 0.001 mol/L $FeSO_4$ solution, 2×10^{-4} mol/L MV solution, 0.1 mol/L Tris-HCl buffer, and 0.004 mol/L H_2O_2 solution were prepared. Then, 0.4 mL of HCBC suspension, 0.4 mL of MV solution, 0.4 mL of $FeSO_4$ solution, 0.4 mL of H_2O_2 solution, and 0.4 mL of Tris-HCl buffer (pH = 3.7) were sequentially added to a beaker. The total volume of the mixture was adjusted to 4 mL with deionized water. After allowing the reaction to proceed for 5 min, the solution was transferred to a cuvette. The absorbance was measured using a UV1750 ultraviolet-visible spectrophotometer (Shimadzu Corporation, Kyoto, Japan), and the hydroxyl radical scavenging rate was calculated based on the absorbance values obtained.

2.5. Boehm Titration Experiment

The acidic oxygen-containing functional groups on the HCBC were quantified via Boehm titration. Briefly, 0.2 g of dried HCBC was placed in a beaker and mixed with 50 mL of a 0.05 mol/L sodium bicarbonate ($NaHCO_3$) solution. The mixture was magnetically stirred at 25 °C for 24 h. After the reaction, the solid was separated by filtration. A 25 mL aliquot of the filtrate was collected and titrated with a standard 0.05 mol/L HCl solution, using 2–3 drops of 0.1% methyl orange as an indicator. The titration endpoint was identified as a color change from yellow to orange, and the volume of HCl consumed was recorded as V_{01}. A blank titration, following the same procedure but in the absence of the HCBC, was performed, and the consumed HCl volume was recorded as V_1. The same procedure was repeated by replacing the $NaHCO_3$ solution with 0.05 mol/L sodium carbonate (Na_2CO_3) and sodium hydroxide (NaOH) solutions, respectively. The corresponding volumes of HCl consumed were recorded as V_{02} and V_{03} for the samples, and V_2 and V_3 for their respective blanks.

The contents of carboxyl groups, lactone groups, and hydroxyl groups in the HCBC were calculated using the following formula.

$$C_{Carboxylic} = \frac{2 \times (V_{o1} - V_1) \times C_{HCl}}{m} \tag{4}$$

$$C_{Lactonic} = \frac{2 \times [(V_{o2} - V_2) - (V_{o1} - V_1)] \times C_{HCl}}{m} \tag{5}$$

$$C_{Hydroxyl} = \frac{2 \times [(V_{o3} - V_3) - (V_{o2} - V_2)] \times C_{HCl}}{m} \tag{6}$$

where C_{HCl} is concentration of standardized HCl (mol/L), and m is mass of the sample (g). The factor "2" accounts for the 25 mL aliquot taken from 50 mL filtrate. All experiments were performed in triplicate, and the average values were reported to ensure reproducibility.

3. Results

3.1. Characterization of HCBCs

3.1.1. FTIR

As shown in Figure 2, the infrared spectroscopy of bentonite exhibits characteristic absorption peaks at 3624 cm^{-1} (Al-OH stretching vibration), 3444 cm^{-1} (adsorbed water OH stretching vibration), 1644 cm^{-1} (adsorbed water -OH bending vibration), 1032 cm^{-1} (Si-O-Si stretching vibration), 521 cm^{-1} (Si-O-Al bending vibration), and 464 cm^{-1} (Si-O-Si bending vibration) [23]. These characteristic peaks remain after the hydrothermal carbonization reaction, and new stretching vibration peaks of C=O and C=C appear at 1704 cm^{-1} and 1631 cm^{-1}, respectively. This indicates that soluble starch is deposited on the surface of bentonite, introducing oxygen-containing functional groups after hydrothermal carbonization [18].

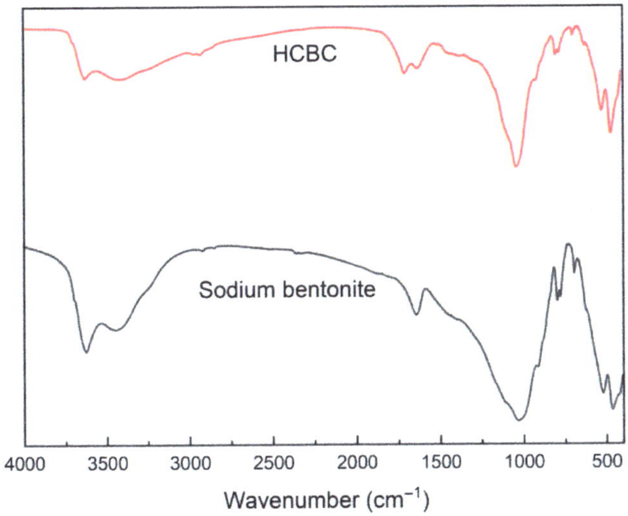

Figure 2. FTIR of sodium bentonite and HCBC.

3.1.2. XPS

From the full-spectrum analysis of bentonite and the HCBC in Figure 3a, a C1s peak appears at a binding energy of 284.8 eV in HCBC, suggesting that the hydrothermal carbonization reaction increases the carbon content on the surface of bentonite. As shown in Figure 3b, the C1s peak is located at 284.8 eV, and its four fitted peaks at 284.8, 285.3, 286.8, and 288.2 eV correspond to aromatic groups or alkyl-substituted aromatic groups (R-C_6H_5), phenolic or ether groups (C_6H_5 O/C-O-C), carbonyl groups (R-C=O), and carboxyl, ester, or lactone groups (-O-C=O), respectively [24]. As shown in Figure 3c, the O1s peak is located at 531.7 eV, and its three fitted peaks at 531.1, 532.0, and 532.7 eV correspond to carbonyl groups (C=O), ester groups (C-O-C), and hydroxyl groups (-OH), respectively [25]. It can be seen that the surface of the composite formed after the hydrothermal carbonization of bentonite contains abundant oxygen-containing functional groups.

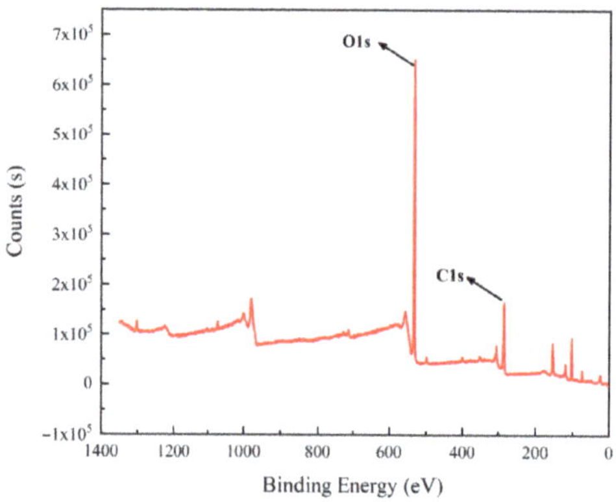

(**a**)

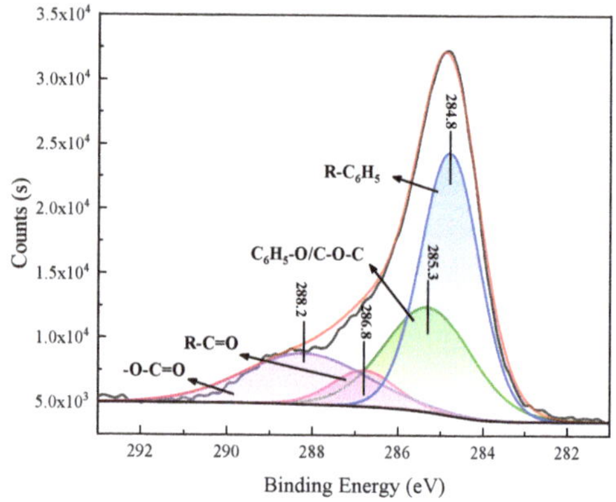

(**b**)

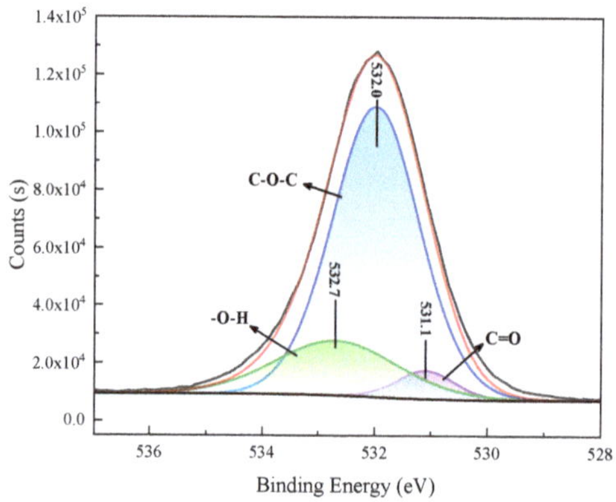

(**c**)

Figure 3. XPS spectra of HCBC: (**a**) survey spectrum, (**b**) C1s spectrum, (**c**) O1s spectrum.

3.1.3. Boehm Titration

Table 1 shows the experimental results of the Boehm titration. The contents of carboxyl groups (-COOH), lactone groups (-O-CO-), and hydroxyl groups (-OH) are 0.5 mmol/g, 0.2 mmol/g, and 0.3 mmol/g, respectively, which is generally consistent with the XPS results. The Boehm titration results further confirm that the surface of the HCBC is rich in oxygen-containing functional groups.

Table 1. Contents of carboxyl, lactone and hydroxyl groups in the HCBC.

Group	Group Content (mmoL·g^{-1})
Carboxyl	0.5
Lactone	0.2
Hydroxyl	0.3

3.1.4. XRD

Figure 4 shows the XRD patterns of bentonite and the HCBC. Compared with bentonite, no new diffraction peaks appeared in the XRD pattern of the HCBC, indicating that the basic crystal structure of bentonite was not destroyed. According to Bragg's equation, $n\lambda = 2d\sin\theta$, where n is 1, the wavelength is 1.54 nm, and the θ for sodium bentonite and the

HCBC is 3.6112° and 3.1573°, detected by XRD. Therefore, the calculated interlayer spacing was 1.22 nm and 1.40 nm for sodium bentonite and the HCBC, respectively. The reason for this is that water-soluble starch was adsorbed into the interlayers of montmorillonite through hydrogen bonds and other interactions. Water-soluble starch would change into carbon particles through hydrothermal carbonization under high temperatures, which could increase the interlayer spacing. Combined with SEM analysis, it can be seen that the hydrothermal carbonization reaction not only occurred on the outer surface of bentonite but also in the interlayer of bentonite.

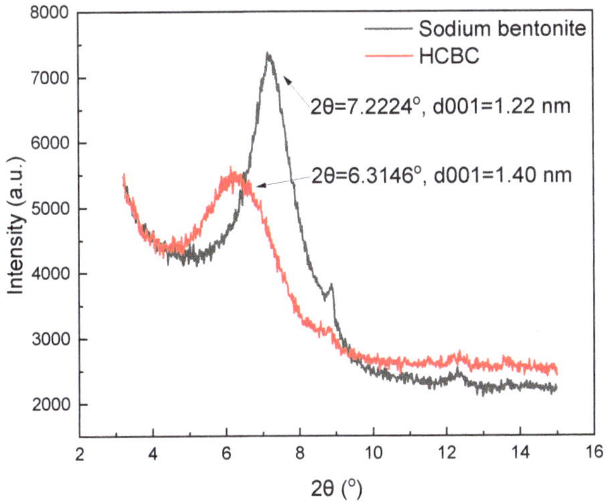

Figure 4. XRD of sodium bentonite and HCBC.

3.1.5. SEM and TEM

It can be seen from the SEM image in Figure 5a that bentonite has a typical plate-like structure. In the HCBC, in addition to retaining some plate-like structure, a large number of nearly spherical and irregular nanoparticles can also be observed, some of which are embedded between the layers of bentonite, while others are in a free deposition state. In a high-temperature hydrothermal environment, starch and other sugar compounds first hydrolyze into low-molecular substances such as glucose, and then dehydrate to form intermediates such as 5-hydroxymethylfurfural. These intermediates undergo condensation or addition reactions to form polyfuran structures, and then undergo aromatization reactions to ultimately form hydrothermal carbon [26]. When bentonite is present, there are two main reaction pathways: a portion of the soluble starch is adsorbed on the surface of bentonite under the influence of hydrogen bonds and the polarity of the bentonite surface, and 5-hydroxymethylfurfural and other intermediates nucleate and grow on the active sites of the bentonite surface to form carbon particles. The growth of these carbon particles is restricted by bentonite, resulting in a smaller particle size, mainly at the nanoscale. Another portion of the soluble starch undergoes hydrothermal carbonization directly in the solution to form micrometer-sized carbon spheres, which deposit on the surface of bentonite [18]. Due to the different reaction pathways, the hydrothermal carbon between the layers of bentonite has a graphite-like structure, while the hydrothermal carbon on the surface of bentonite has an amorphous structure [27]. After the surface of bentonite is covered with micro–nano carbon spheres, the BET specific surface area decreases from 6.7379 m^2/g to 2.8357 m^2/g. It can also be seen from the TEM images in Figure 5c,d that a typical plate-like aggregation structure can be observed in the bentonite suspension, while in the

HCBC suspension, not only are there a large number of nearly spherical nanoparticles deposited on the surface of the bentonite layers, but also hundreds of nanometer-sized or even micrometer-sized free particles can be observed.

Figure 5. SEM and TEM images of HCBC and bentonite. (**a**) SEM image of bentonite, (**b**) SEM image of HCBC, (**c**) TEM image of bentonite, and (**d**) TEM image of HCBC.

3.1.6. TGA

As shown in Figure 6, for bentonite, the mass loss between 30 and 200 °C mainly results from the removal of free water and bound water, and the mass loss between 400 and 600 °C corresponds to the dehydroxylation reaction. When the temperature rises to 1000 °C, the mass loss rate is 11.47%. For the HCBC, starch molecules enter the interlayer of bentonite and undergo hydrothermal carbonization reactions, displacing some water molecules. Within the range of 30 to 200 °C, the weight loss is lower than that of bentonite. After exceeding 350 °C, HCBC begins to lose weight rapidly, and at this time, the carbon particles loaded on the surface of bentonite start to degrade. When the temperature rises to 1000 °C, the mass loss rate is 24.28%. From this, it can be inferred that the content of hydrothermal carbon particles loaded on the surface of bentonite is approximately 12.81%.

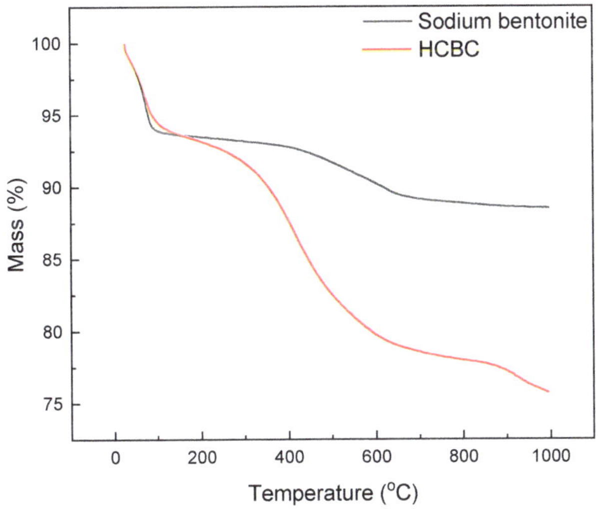

Figure 6. TGA curves of sodium bentonite and the HCBC.

3.1.7. Particle Size Distribution and Water Adsorption

As depicted in Figure 7a, both sodium bentonite and the HCBC exhibit multimodal distribution. The average particle size of sodium bentonite and the HCBC is 5.434 μm and 0.811 μm, respectively, indicating that after modification, the dispersion of bentonite particles is significantly improved. This could also be verified by the zeta potential measurement results. The zeta potential of bentonite being −38.41 mV changes to −45.51 mV for the HCBC. As shown in Figure 7b, for bentonite, the weight increases rapidly at the initial 2 h, followed by a gradual increase. After about 56.5 h, the weight increase is 33%, which corresponds to the amount of water adsorption. In terms of the HCBC, the weight increases rapidly at the initial 1.3 h, then increases with a much slower rate. After 56.5 h, the weight increases only by 14%, much lower than that of bentonite. This indicates that the water affinity is decreased to some degree after modification with the hydrothermal carbon clusters on the bentonite surface.

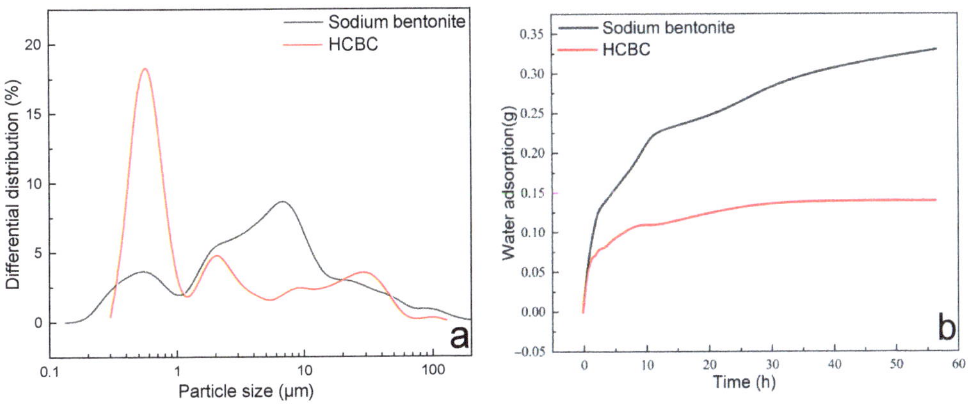

Figure 7. Particle size distribution (**a**) and water adsorption amount (**b**) as a function of time for bentonite and HCBC.

3.2. Properties Evaluation

3.2.1. Xanthan Slurries

The variation in rheological parameters including PV, YP, gel strength, filtration loss, and extreme pressure lubrication coefficient of XC slurries as a function of HCBC concentration before and after hot rolling (AHR) at 120 °C is presented in Figure 8. Before hot rolling (BHR), the rheological parameters such as PV, YP, and filtration loss changed slightly with the increasing concentration of HCBC, indicating that the HCBC imposes little influence on the rheological and filtration properties of XC slurries. After hot rolling at 120 °C, the degradation of XC resulted in remarkable decrease in PV and YP for the control slurry, while the addition of HCBC was able to restore the PV to that of before hot rolling, and increase the YP to some extent. Meanwhile, both the filtration loss and lubrication coefficient decreased obviously with the addition of HCBC. A reduction of 55.4% and 30.3% was observed, respectively, for the filtration loss and lubrication coefficient at 2 w/v% HCBC. In terms of gel strength, as shown in Figure 8c,d, both the gel 10s and gel 10 min decreased with the increase in HCBC concentration before hot rolling. After hot rolling, these two parameters increased progressively with the increase in HCBC concentration, indicating the improved suspending capacity. As depicted in Figure 9, after hot rolling, the AV retention rate was only 23.7% for the XC slurries, indicating the severe degradation of XC. However, it recovered to over 50% when the concentration of HCBC was above 0.5%. Howard et al. (2015) stated that the fluids maintaining 50% of their viscosity after hot rolling for 16 h can be defined to have thermal stability [28]. Therefore, the addition of HCBC improves the thermal stability of XC slurries effectively. Overall, the XC slurries with HCBC maintain stable viscosity, low filtration loss, and low lubrication coefficient after hot rolling.

3.2.2. Modified Starch Slurries

The variation in rheological parameters including PV, YP, gel strength, filtration loss, and extreme pressure lubrication coefficient of modified starch slurries as a function of HCBC concentration before and after hot rolling at 150 °C is presented in Figure 10. Regarding the rheological parameters, the PV exhibited an increasing trend with the addition of HCBC before hot rolling and after hot rolling, whereas the YP generally increased before hot rolling but decreased slightly after hot rolling with the addition of HCBC. The gel 10s decreased gradually, and gel 10 min exhibited an initial decrease and then a slight increase behavior with the increase in HCBC concentration before hot rolling. After hot rolling, both the gel 10 s and gel 10 min exhibited an increase after the incorporation of HCBC, indicating enhanced suspending capacity. In terms of filtration and lubrication, they exhibited a similar pattern of change as XC slurries. When 3 w/v% HCBC was employed, the filtration loss and lubrication coefficient was reduced by 67.2% and 20.5%, respectively. For the AV retention rate, as shown in Figure 11, it increased at a low concentration of HCBC, while decreasing slightly at relatively high concentration of HCBC. Nevertheless, for all the concentrations of HCBC, the AV retention rate is higher than 50%, indicating thermal stability after thermal treatment.

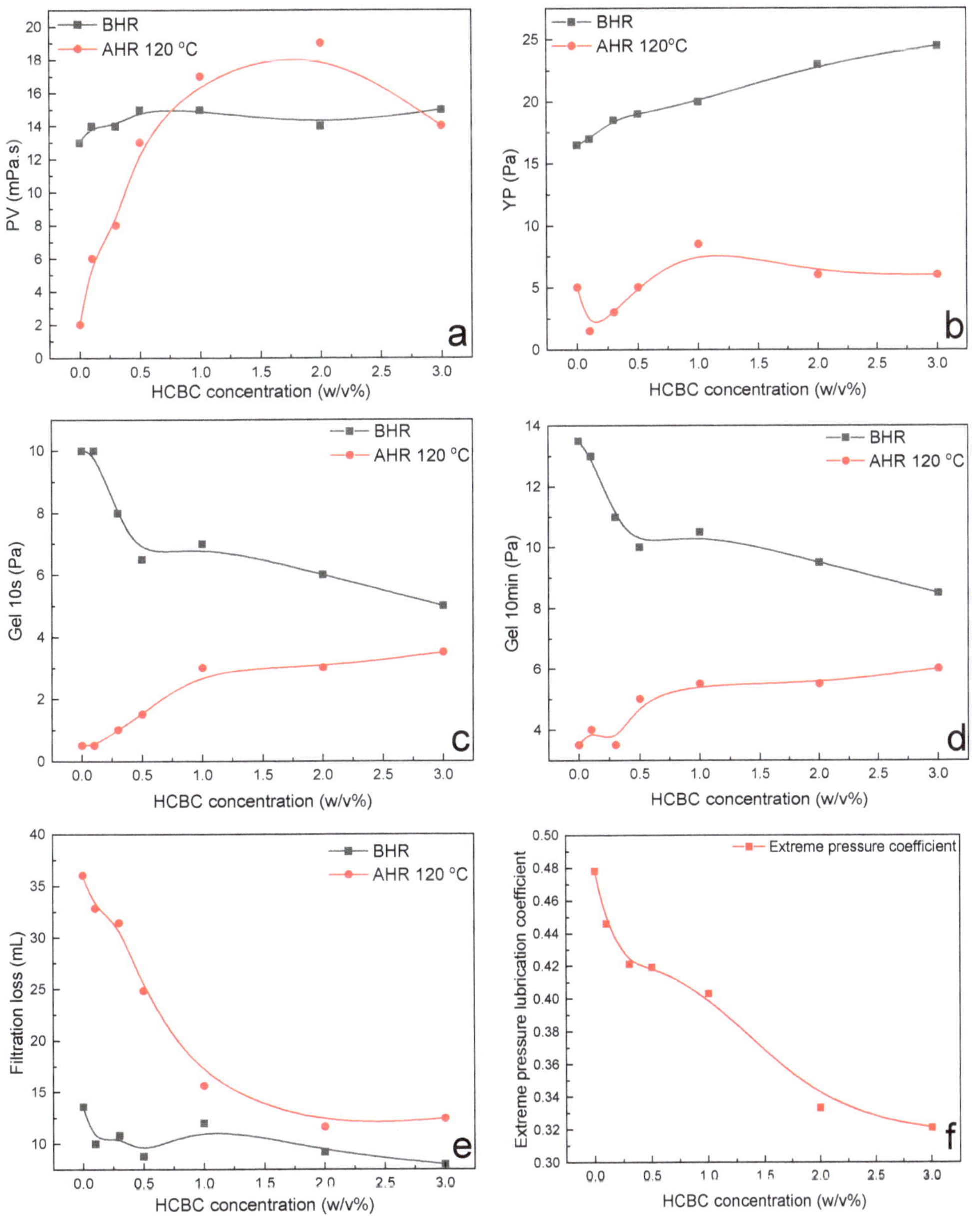

Figure 8. Variation in rheological, filtration, and lubrication properties of XC slurry as a function of HCBC concentration before and after hot rolling at 120 °C: (**a**) PV, (**b**) YP, (**c**) Gel 10s, (**d**) Gel 10min, (**e**) filtration loss, (**f**) extreme pressure lubrication coefficient.

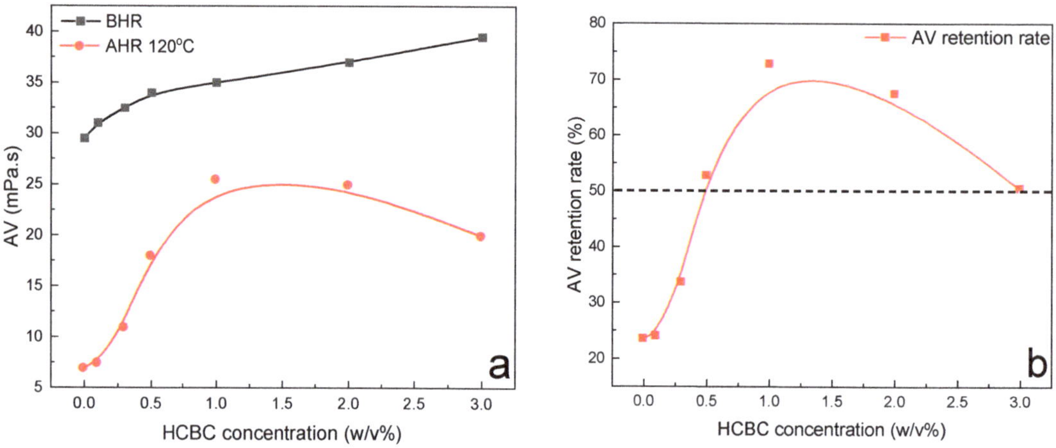

Figure 9. Variation in AV (**a**) and AV retention rate (**b**) as a function of HCBC concentration for the XC slurry after hot rolling at 120 °C.

3.2.3. HT Polymer Slurries

The variation in rheological parameters including PV, YP and gel strength, filtration loss, and extreme pressure lubrication coefficient of HT polymer slurries as a function of HCBC concentration before and after hot rolling at 200 °C is presented in Figure 12. Since HCBC are microparticles, the incorporation of HCBC increases the friction between the liquid and solid phase, as well as the friction between solid and solid phase, resulting in the increase in PV to some extent before hot rolling. The oxygen-containing functional groups in HCBC probably promoted the formation of network structures among bentonite particles, HT polymers, and HCBC through hydrogen bonding. The YP increased obviously at 3.0 w/v% HCBC before hot rolling. After hot rolling at 200 °C, both PV and YP for the control slurries decreased to near zero, indicating that the slurries suffering from severe degradation under such harsh conditions. However, the addition of HCBC with much low concentration of 0.3 w/v% can maintain the PV and YP value approaching to that of before hot rolling. For gel strength, both gel 10 s and gel 10 min decreased at low concentration of HCBC and then increased to some degree before hot rolling. After hot rolling, they all exhibited a similar behavior that increased obviously at low concentration of HCBC and then varied slightly. Overall, the addition of HCBC could improve the gel structure of the slurries after hot rolling. As shown in Figure 13, the AV retention rate was even higher than 100% when the concentration of HCBC above 0.5 w/v%, indicating excellent thermal stability effect. For filtration loss and lubrication, the addition of 2 w/v% HCBC resulted in decrease by 70.4% and 77.5%, respectively, after hot rolling at 200 °C.

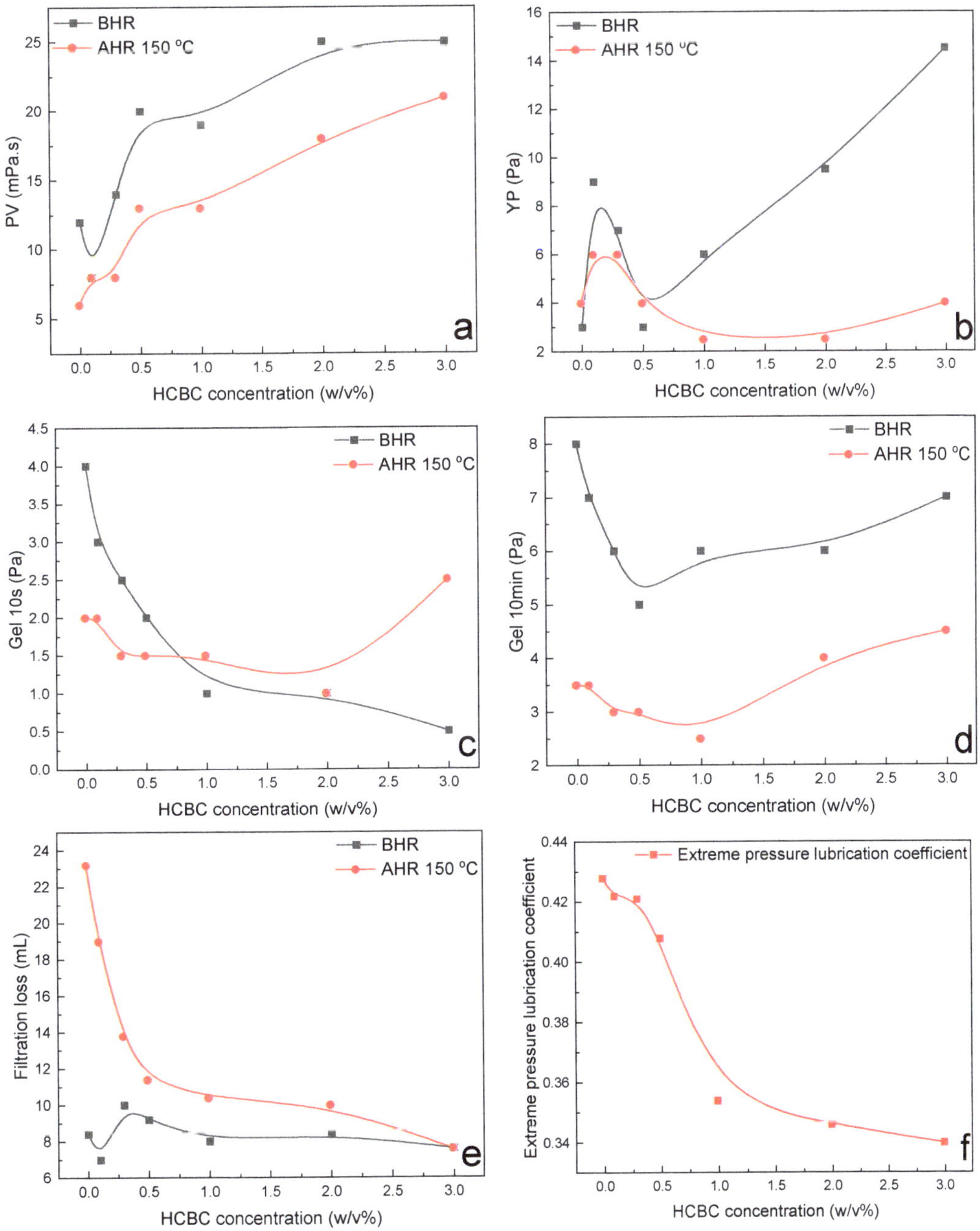

Figure 10. Variation in rheological, filtration, and lubrication properties of modified starch slurry as a function of HCBC concentration before and after hot rolling at 150 °C: (**a**) PV, (**b**) YP, (**c**) Gel 10s, (**d**) Gel 10min, (**e**) filtration loss, (**f**) extreme pressure coefficient.

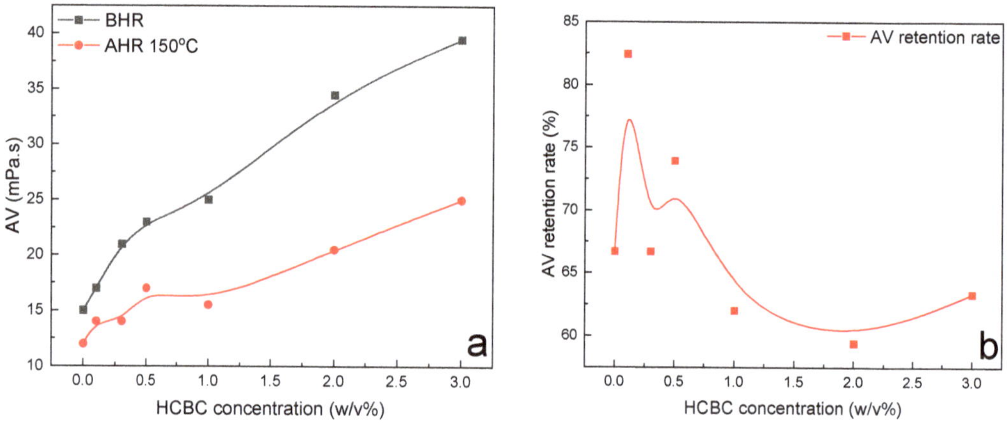

Figure 11. Variation in AV (**a**) and AV retention rate (**b**) as a function of HCBC concentration for the modified starch slurry after hot rolling at 150 °C.

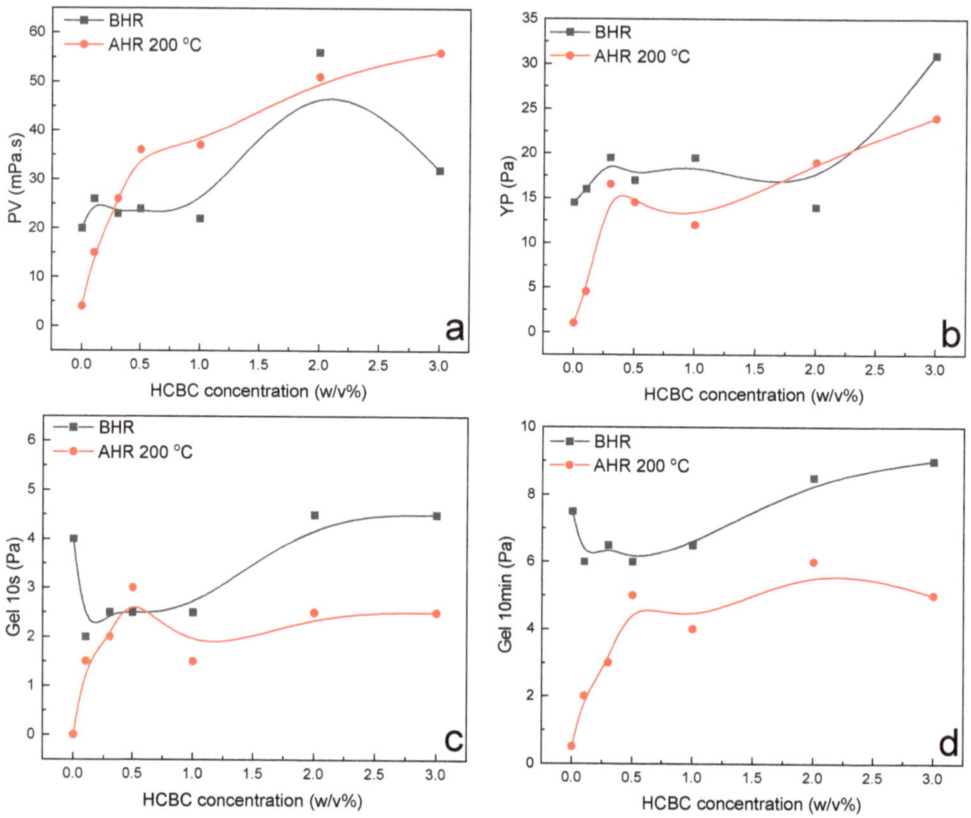

Figure 12. *Cont.*

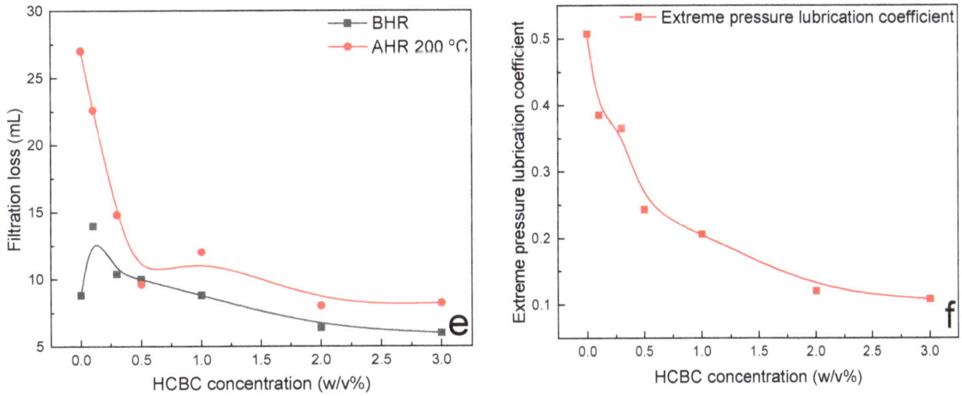

Figure 12. Variation in rheological, filtration, and lubrication properties of HT polymer slurry as a function of HCBC concentration before and after hot rolling at 200 °C: (**a**) PV, (**b**) YP, (**c**) Gel 10s, (**d**) Gel 10min, (**e**) filtration loss, (**f**) extreme pressure coefficient.

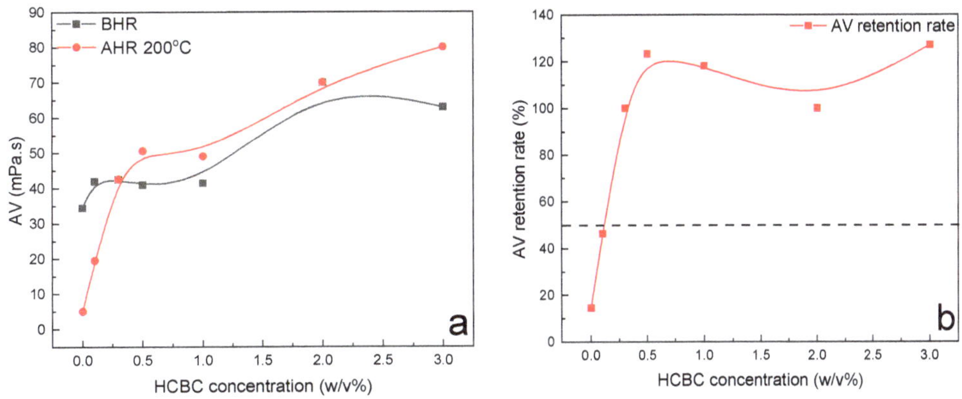

Figure 13. Variation in AV (**a**) and AV retention rate (**b**) as a function of HCBC concentration for the HT polymer slurry after hot rolling at 200 °C.

3.2.4. Free Radical Scavenging

The variation in hydroxyl radical scavenging rate with HCBC concentration was calculated based on the change in maximum absorbance, as shown in Figure 14. At an extremely low concentration of 0.025 mg/mL, HCBC achieves a scavenging rate of 5.39%. Thereafter, the scavenging rate generally shows an upward trend with increasing concentration. When the concentration reaches 0.625 mg/mL, the scavenging rate can reach 67.35%, indicating that HCBC effectively blocks the reaction between hydroxyl radicals and methyl violet, and that it exhibits a scavenging effect on hydroxyl radicals. After the radicals generated through chain initiation are scavenged, the radical chain propagation reaction is interrupted, preventing subsequent oxidation reactions. This can effectively avoid the attack and damage of free radicals on polymer molecular chains.

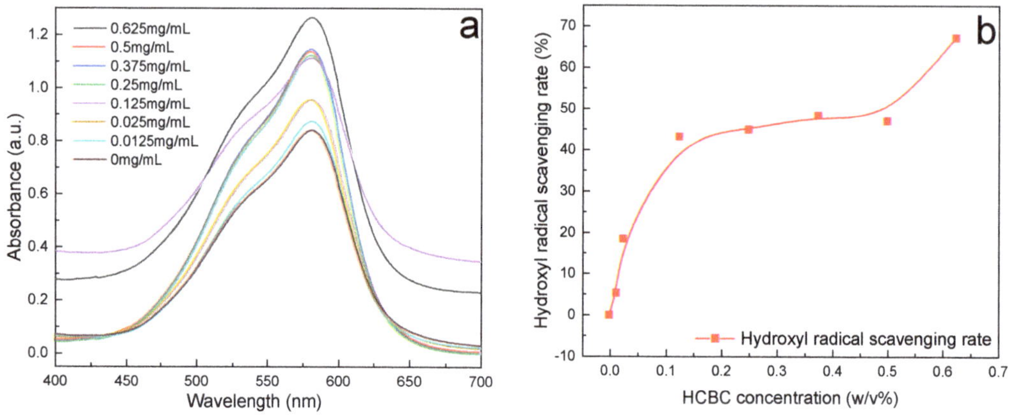

Figure 14. Hydroxyl radical scavenging as a function of HCBC concentration: (**a**) absorbance; (**b**) hydroxyl radical scavenging rate.

3.2.5. Filtration Loss

The LTLP filtration loss and HTHP filtration loss of bentonite slurry treated with various filtration reducers are illustrated in Figure 15. Before hot rolling, the addition of HCBC decreased the LTLP filtration loss from 20 mL to 17.2 mL, exhibiting a limited effectiveness, whereas the HT polymer could decrease the filtration loss from 20 mL to 10 mL, better than HCBC. After hot rolling at 220 °C, due to the degradation of the HT polymer, the fluid with the HT polymer lost filtration control, with a high filtration loss of 37 mL. The combination of the HCBC and the HT polymer exhibited much lower filtration loss, indicating a synergistic effect. As shown in Figure 13b, after hot rolling, the dehydration of bentonite resulted in the aggregation of clay particles, which corresponded to the quite high HTHP (200 °C/3.5 MPa) filtration loss of 640 mL. The degradation of the HT polymer also caused uncontrollable HTHP filtration loss, and the HTHP filtration loss reached 620 mL. The combination of 1 w/v% HT polymer and 1 w/v% bentonite decreased the HTHP filtration loss to a very limited degree. However, the HTHP filtration loss was significantly decreased to 180 mL in the presence of 1 w/v% HCBC, and further decreased to 84 mL when 1 w/v% HCBC and 1 w/v% HT polymer were both used. Both the LTLP filtration loss and the HTHP filtration loss indicated that the HCBC can effectively decrease the filtration loss of bentonite slurry and has a synergistic effect with HT polymer.

The SEM images of filter cake formed by bentonite slurry with and without HCBC before and after hot rolling are shown in Figure 16. Before hot rolling, the clay particles are fully hydrated and dispersed, and the particles are mainly connected by end to surface to form a typical honeycomb structure. Due to the thick hydration film, the edges of clay particles are relatively rounded after rapid freezing and freeze-drying with liquid nitrogen. After hot rolling at 220 °C, as shown in Figure 16b, due to the high-temperature dehydration effect, the repulsive force of the hydration film decreases, and the clay particles form larger sheet-like structures through surface to surface connections. The filtration channels significantly increase and the filtration loss significantly increases.

After adding HCBCs, as shown in Figure 16c, the oxygen-containing functional groups on the surface of HCBCs interact with clay particles before hot rolling, causing the clay particles to stick together and form sheets, which to some extent reduces the filtration area. As shown in Figure 16d, after hot rolling at 220 °C, the clay particles in the filter cake with added HCBCs also underwent agglomeration due to high-temperature dehydration. However, HCBCs enhanced the repulsive force between clay particles, and the agglomeration effect was significantly weakened compared to the filter cake without HCBCs. In addition, regardless of whether hot rolling has been carried out or not, a large number of micro and nano carbon spheres can be observed in the filter cake after adding HCBCs, which is beneficial for improving particle stacking efficiency and forming a dense mud cake.

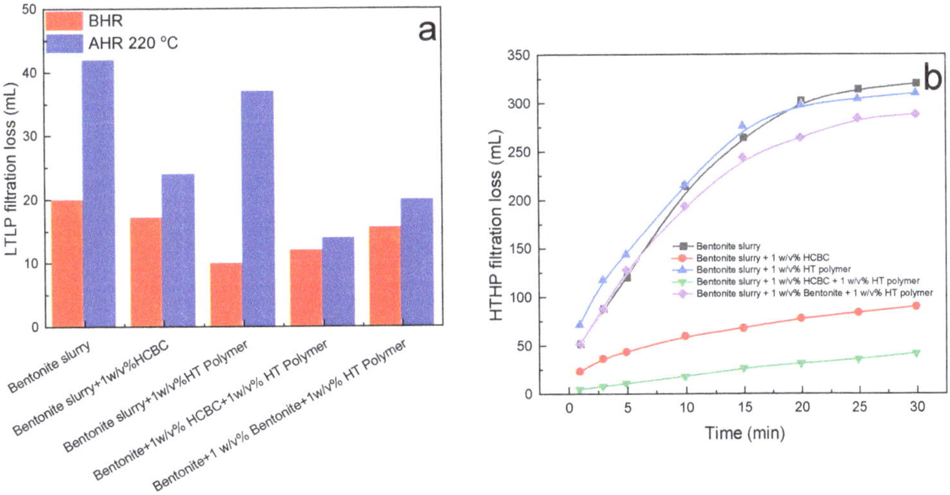

Figure 15. The LTLP filtration loss (**a**) and HTHP filtration loss (**b**) of bentonite slurry in the presence of various filtration reducers before and after hot rolling at 220 °C.

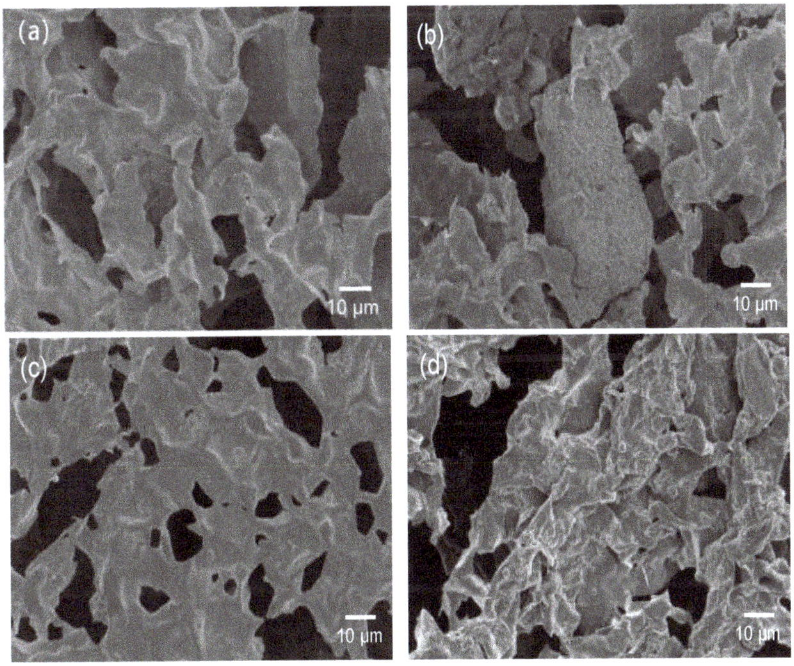

Figure 16. SEM images of filter cake: (**a**) bentonite slurry before hot rolling, (**b**) bentonite slurry after hot rolling at 220 °C, (**c**) bentonite slurry in the presence of HCBC before hot rolling, (**d**) bentonite slurry in the presence of HCBC after hot rolling at 220 °C.

3.3. High Temperature Stabilizing and Filtration Control Mechanism

When starch is dissolved in water under subcritical conditions, it first hydrolyze into low molecular weight mono saccharides such as glucose, and then dehydrate to form intermediates such as 5-hydroxymethylfurfural. The intermediates undergo condensation or addition reactions to form polyfuran structures, which are then subjected to aromatization reactions to ultimately generate hydrothermal carbon.

When bentonite is present, there are two main reaction pathways: a portion of soluble starch is adsorbed on the surface of bentonite under hydrogen bonding and polarity induction. Intermediates such as 5-hydroxymethylfurfural nucleate and grow at the active sties on the surface of bentonite to form carbon particles. The growth of this part of carbon particles is limited by bentonite; therefore, their particle size is relatively small, mainly at the nanoscale. For the other part of soluble starch, which is dissolved in the aqueous solution, it can directly form micrometer-sized carbon spheres through hydrothermal carbonization. These carbon spheres can deposit on the surface of bentonite.

The anchoring or depositing of hydrothermal carbon spheres on the surface of bentonite brings abundant oxygen-containing groups such as hydroxyl, carbonyl, and carboxyl groups. On the one hand, the hydroxyl groups tend to act as H-atom donors to unstable free radical molecules [29]. On the other hand, compounds with long conjugated C=C chains are usually great free radical scavengers. For graphene, carbon nanotubes, and fullerene, the radical addition to the sp^2 carbon network plays an important role [30]. Similarly, there are plentiful sp^2-conjugated C=C chains in the core of hydrothermal carbon spheres. The spin across the conjugated graphemic backbone is delocalized, which forms free radical adducts and results in the decrease in free radical quantities [31,32]. The combination of hydrogen donating by surface hydroxyl groups and formation adducts by the core sp^2 C=C carbon contributes to the free scavenging effect, which in turn prevents the thermal oxidative degradation of polymers. Therefore, the thermal stability of water-based drilling fluid is significantly enhanced.

Regarding filtration loss, due to the abundant oxygen groups on the HCBC, on the one hand, the addition of HCBC promotes the dispersibility of clay particles. On the other hand, the partially free nano carbon spheres in the HCBC increase the content of submicron particles in the system. Therefore, the addition of HCBC leads to a decrease in the average particle size of the suspension, and forms a reasonable gradation with bentonite particles, making it easier to form a dense filter cake and thus reducing filtration loss. At the same time, clay particles also undergo agglomeration due to high-temperature dehydration, but the HCBC enhances the repulsion between clay particles, and the agglomeration is significantly weakened. In addition, regardless of before and after hot rolling, a large number of micro and nano carbon spheres are observed to be filled into the filter cake after adding the HCBC, which is beneficial for improving particle packing efficiency and forming a dense filter cake.

4. Conclusions

In this study, hydrothermal carbon/bentonite composites (HCBCs) were prepared by a hydrothermal carbonization reaction using soluble starch and sodium bentonite as raw materials. The interlayer spacing of bentonite increased from 1.22 nm to 1.40 nm after the hydrothermal carbonization reaction. There were plentiful micro- and nano-sized carbon spheres deposited on the surface of bentonite, and the oxygenated groups on the carbon sphere surface improved the dispersion stability of bentonite particles.

HCBCs exhibit limited effect on the rheology of xanthan slurries, modified starch slurries, and high-temperature-resistant polymer slurries. However, after dynamical thermal aging, the presence of HCBCs could effectively improve the apparent viscosity retention, reduce the filtration loss, and enhance the lubrication of the slurries.

The high efficiency in free radical scavenging contributes to the excellent thermal stability of water-based drilling fluids. The surface of HCBCs has abundant oxygen-containing functional groups, which improve the dispersion stability of clay particles at ultra-high temperatures through electrostatic repulsion and other mechanisms. The relatively small particle size of HCBC and the formation of free micro nano carbon spheres through hydrothermal reactions are beneficial for improving the solid-phase particle size distribution of drilling fluids and enhancing the quality of mud cakes. The above comprehensive effects mean that HCBCs exhibit excellent ultra-high temperature filtration performance.

The raw materials used in the preparation of HCBC are widely sourced, and the preparation process is green and environmentally friendly. Multifunctional HCBCs show great potential in developing high-performance water-based drilling fluids. This study also opens up a new way to design and develop multifunctional and environmentally friendly additives for drilling fluids.

Author Contributions: Conceptualization, H.Z. and D.L.; methodology, H.Z. and Y.Z.; validation, C.C. and Y.Z.; investigation, H.Z.; resources, H.Z. and X.W.; data curation, H.Z. and C.C.; writing—original draft preparation, H.Z.; writing—review and editing, H.Z. All authors have read and agreed to the published version of the manuscript.

Funding: This research was funded by National Science and Technology Major Projects, grant number 2025ZD1401301, the Open Fund for Sinopec's Key Laboratory of Ultra-Deep Well Drilling Engineering and Technology, and the National Natural Science Foundation of China, grant number No. 52174013.

Data Availability Statement: Data is unavailable due to privacy.

Conflicts of Interest: Author Yubin Zhang, Daqi Li, and Xianguang Wang was employed by SINOPEC Research Institute of Petroleum Engineering Co., Ltd. The remaining authors declare that the research was conducted in the absence of any commercial or financial relationships that could be construed as a potential conflict of interest.

Abbreviations

The following abbreviations are used in this manuscript:

AV	Apparent viscosity
FTIR	Fourier Transform infrared spectroscopy
HCBCs	Hydrothermal carbon/bentonite composites
HT	High-temperature
HTHP	High-temperature and high-pressure
LTLP	Low-temperature and low-pressure
MV	Methyl violet
PV	Plastic viscosity
SEM	Scanning Electron Microscope
TEM	Transmission electron microscope
TGA	Thermogravimetry Analysis
XC	Xanthan gum
XPS	X-ray photoelectron spectroscopy
YP	Yield point
BHR	Before hot rolling
AHR	After hot rolling

References

1. Galindo, K.A.; Zha, W.; Zhou, H.; Deville, J.P. Clay-free high performance water-based drilling fluid for extreme high temperature wells; SPE-173017-MS. In Proceedings of the SPE/IADC Drilling Conference and Exhibition, London, UK, 17–19 March 2015.
2. Tehrani, M.A.; Popplestone, A.; Guarneri, A.; Carminati, S. Water-based drilling fluid for HT/HP applications; SPE-105485-MS. In Proceedings of the International Symposium on Oilfield Chemistry, Houston, TX, USA, 28 February–2 March 2007.
3. Rabaioli, M.R.; Miano, F.; Lockhart, T.P.; Burrafato, G. Physical/chemical studies on the surface interactions of bentonite with polymeric dispersing agents. SPE-25179-MS. In Proceedings of the SPE International Symposium on Oilfield Chemistry, New Orleans, LA, USA, 2–5 March 1993.
4. Ahmad, H.M.; Kamal, M.S.; Al-Harthi, M.A. Effect of thermal aging and electrolyte on bentonite dispersions: Rheology and morphological properties. *J. Mol. Liq.* **2018**, *269*, 278–286. [CrossRef]
5. Cui, J.; Zhang, Z.; Han, F. Effects of pH on the gel properties of montmorillonite, palygorskite and montmorillonite-palygorskite composite clay. *Appl. Clay Sci.* **2020**, *190*, 105543. [CrossRef]
6. Dormán, J. Chemistry and field practice of high-temperature drilling fluids in Hungary; SPE-21940-MS. In Proceedings of the SPE/IADC Drilling Conference, Amsterdam, The Netherlands, 11–14 March 1991.
7. Elward-Berry, J.; Darby, J.B. Rheologically sable, nontoxic, high-temperature, water-based drilling fluid. *SPE Drill. Complet.* **1997**, *12*, 158–162. [CrossRef]
8. Zilch, H.E.; Otto, M.J.; Pye, D.S. The evolution of geothermal drilling fluid in the imperial valley; SPE 21786. In Proceedings of the Western Regional Meeting, Long Beach, CA, USA, 20–22 March 1991.
9. Lozano, J.; Miska, N.; Takach, N.; Yu, M.; Saasen, A. The effect of elongational flow through the drill bit on the rheology of polymeric drilling fluids; SPE-99107-MS. In Proceedings of the IADC/SPE Drilling Conference, Miami, FL, USA, 21–23 February 2006.
10. Zhou, H.; Deville, J.P.; Davis, C.L. Novel thermally stable high-density brine-based drill-in fluids for HP/HT applications; SPE-172659-MS. In Proceedings of the SPE Middle East Oil & Gas Show and Conference, Manama, Bahrain, 8–11 March 2015.
11. Yang, J.; Sun, J.; Wang, R.; Liu, F.; Wang, J.; Qu, Y.; Wang, P.; Huang, H.; Liu, L.; Zhao, Z. Laponite-polymer composite as a rheology modifier and filtration loss reducer for water-based drilling fluids at high temperature. *Colloids Surf. A Physicochem. Eng. Asp.* **2022**, *655*, 130261. [CrossRef]
12. Ahmed, A.; Pervaiz, E.; Abdullah, U.; Noor, T. Optimization of water based drilling fluid properties with the SiO_2/g-C_3N_4 hybrid. *ACS Omega* **2024**, *9*, 15052–15064.
13. Ao, T.; Yang, L.; Xie, C.; Jiang, G.; Wang, G.; Liu, Z.; He, X. Zwitterionic silica-based hybrid nanoparticles for filtration control in oil drilling conditions. *ACS Appl. Nano Mater.* **2021**, *4*, 11052–11062. [CrossRef]
14. Ikram, R.; Jan, B.M.; Vejpravova, J. Towards recent tendencies in drilling fluids: Application of carbon-based nanomaterials. *J. Mater. Res. Technol.* **2021**, *15*, 3733–3758. [CrossRef]
15. Ospanov, Y.K.; Kudaikulova, G.A. A comprehensive review of carbon nanomaterials in the drilling industry. *J. Polym. Sci.* **2024**, 1–20. [CrossRef]
16. Rana, A.; Khan, I.; Saleh, T.A. Advances in carbon nanostructures and nanocellulose as additives for efficient drilling fluids: Trends and future perspective-A review. *Energy Fuels* **2021**, *35*, 7319–7339. [CrossRef]
17. Li, T.; Shen, J.; Huang, S.; Li, N.; Ye, M. Hydrothermal carbonization synthesis of a novel montmorillonite supported carbon nanosphere adsorbent for removal of Cr (VI) from waste water. *Appl. Clay Sci.* **2014**, *93–94*, 48–55. [CrossRef]
18. Liu, C.; Cai, W.; Liu, L. Hydrothermal carbonization synthesis of Al-pillared montmorillonite@ carbon composites as high performing toluene adsorbents. *Appl. Clay Sci.* **2018**, *162*, 113–120. [CrossRef]
19. Tian, S.; Liu, Y.; Liu, S.; Zeng, G.; Jiang, L.; Tan, X.; Huang, X.; Yin, Z.; Liu, N.; Li, J. Hydrothermal synthesis of montmorillonite/hydrochar nanocomposites and application for 17β-estradiol and 17α-ethynylestradiol removal. *RSC Adv.* **2018**, *8*, 4273–4283. [CrossRef]
20. *API RP 13B-1*; Recommended Practice for field Testing Water-Based Drilling Fluids. American Petroleum Institute: Washington, DC, USA, 2019.
21. Lankone, R.S.; Deline, A.R.; Barclay, M.; Fairbrother, D.H. UV-Vis quantification of hydroxyl radical concentration and dose using principal component analysis. *Talanta* **2020**, *218*, 121148. [CrossRef] [PubMed]
22. Xue, Y.; Luan, Q.; Yang, D.; Yao, X.; Zhou, K. Direct evidence for hydroxyl radical scavenging activity of Cerium oxide nanoparticles. *J. Phys. Chem.* **2011**, *115*, 4433–4438. [CrossRef]
23. Ai, L.; Li, L. Efficient removal of organic dyes from aqueous solution with ecofriendly biomass-derived carbon@montmroillonite nanocomposites by one-step hydrothermal process. *Chem. Eng. J.* **2013**, *223*, 688–695. [CrossRef]
24. Sevilla, M.; Fuertes, A.B. Chemical and structural properties of carbonaceous products obtained by hydrothermal carbonization of saccharides. *Chem. Eur. J.* **2009**, *15*, 4195–4203. [CrossRef]
25. Li, M.; Li, W.; Liu, S.X. Hydrothermal synthesis, characterization, and KOH activation of carbon spheres from glucose. *Carbohydr. Res.* **2011**, *346*, 999–1004. [CrossRef]

26. Titirici, M.M.; Antonietti, M.; Baccile, N. Hydrothermal carbon from biomass: A comparison of the local structure from poly- to monosaccharides and pentoses/hexoses. *Green Chem.* **2008**, *10*, 1204–1212. [CrossRef]
27. Yin, J.L.; Pei, M.S.; He, Y.J.; Du, Y.; Guo, W.; Wang, L. Hydrothermal and activated synthesis of adsorbent montmorillonite supported porous carbon nanospheres for removal of methylene blue from waste water. *RSC Adv.* **2015**, *5*, 89839–89847. [CrossRef]
28. Howard, S.; Kaminski, L.; Downs, J. Xanthan stability in formate brines-formulating non-damaging fluids for high temperature applications; SPE-174228-MS. In Proceedings of the SPE European Formation Damage Conference and Exhibition, Budapest, Hungary, 3–5 June 2015.
29. Wang, Y.; Kong, W.; Wang, L.; Zhang, J.Z.; Li, Y.; Liu, X.; Li, Y. Optimizing oxygen functional groups in graphene quantum dots for improved antioxidant mechanism. *Phys. Chem. Chem. Phys.* **2019**, *21*, 1336–1343. [CrossRef]
30. Li, Q.; Shen, X.; Xing, D. Carbon quantum dots as ROS-generator and -scavenger: A comprehensive review. *Dye. Pigment.* **2023**, *208*, 110784. [CrossRef]
31. Innocenzi, P.; Stagi, L. Carbon dots as oxidant-antioxidant nanomaterials, understanding the structure-properties relationship. A critical review. *Nanotoday* **2023**, *50*, 101837. [CrossRef]
32. Qiu, Y.; Wang, Z.; Owens, A.C.E.; Kulaots, I.; Chen, Y.; Kane, A.B.; Hurt, R.H. Antioxidant chemistry of graphene-based materials and its role in oxidation protection technology. *Nanoscale* **2014**, *6*, 11744–11755. [CrossRef]

minerals MDPI

Article

Mineral Characterization and High Resistivity Analysis of Ultra-Deep Shale from Mahu Sag, China

Yangfei Yu [1,2], Ding Zhang [1], Panpan Zhang [1,3], Zongjie Mu [1,3,*], Shouceng Tian [1], Yawen Tan [1] and Ronghao Zhou [1]

1 The Karamay Branch of State Key Laboratory of Petroleum Resources and Prospecting, China University of Petroleum (Beijing) at Karamay, Karamay 834000, China
2 Xinjiang Oilfield Company, PetroChina, Karamay 834000, China
3 Xinjiang Key Laboratory of Intelligent Petroleum Exploration and Engineering, Karamay 834000, China
* Correspondence: muzongjie@cupk.edu.cn

Abstract

Ultra-deep shale in the Mahu Sag, characterized by difficult-to-drill formations, exhibits high resistivity. This study uses XRD and petrophysical testing on 12 dry core samples (depths 4600–5000 m) to characterize mineral composition and evaluate resistivity-influencing factors. Mineralogical analysis reveals that brittle minerals, dominated by quartz and feldspar (>50%), constitute the primary components of the ultra-deep shale in the Mahu Sag, with quartz, feldspar, and carbonates collectively accounting for ~80%. Clay (~6%) and pyrite (<5%) contents are notably low, resulting in elevated resistivities of 10^5–10^7 Ω·m. Resistivity correlates negatively with pyrite (r = −0.588) and feldspar (r = −0.319) but positively with dolomite (r = 0.209), quartz (r = 0.017), and porosity (r = 0.749). At elevated temperatures (100 °C), resistivity declines owing to enhanced ionic conduction. These findings clarify high-resistivity mechanisms, supporting resistivity-based drilling parameter optimization.

Keywords: ultra-deep shale; mineral characterization; high resistivity; drillability prediction

1. Introduction

Shale oil, as a crucial unconventional hydrocarbon resource, has emerged as a research hotspot in the global energy sector in recent years. Compared to conventional reservoirs, shale oil exhibits widespread resource distribution, ultra-low permeability, and complex accumulation mechanisms [1]. China's continental shale oil resources are abundant and widely distributed, with geological resources exceeding 300 billion tons [2–4]. As exploration and development targets shift from shallow to ultra-deep layers, proven reserves of ultra-deep shale oil continue to increase [5]. Currently, the deepest target formation for continental shale oil in China is the Permian Fengcheng Formation in the Mahu Sag of the Junggar Basin, with burial depths ranging from 4500 to 6100 m and potential resources of approximately 530 million tons [6–8]. The ultra-deep shale in the Mahu Sag exhibits substantial potential for large-scale hydrocarbon production. Efficient exploitation of such reservoirs advances theoretical models for unconventional resource development and offers practical guidance for global ultra-deep shale oil exploration.

The ultra-deep continental shale in the Mahu Sag has undergone multiple tectonic movements and diagenetic processes, exhibiting characteristics of high temperature (>110 °C), high geostress (>80 MPa), high rock strength (~500 MPa), and strong heterogeneity ("three highs and one strong") [9]. High geostress and rock strength substantially

 https://doi.org/10.3390/min15111171

elevate the energy demands for rock resistance during cutting and crushing. Meanwhile, pronounced heterogeneity induces uneven stress distribution and hinders local fracturing. These factors collectively intensify bit wear, diminish energy transfer efficiency, and heighten rock-breaking difficulty. Conventional drilling speed-up measures are ineffective, and low rock-breaking efficiency has become a core bottleneck constraining the economic development of ultra-deep continental shale oil in the Mahu Sag. Traditional empirical-based drilling parameter adjustment models can no longer meet current speed-up demands.

Resistivity, as a key parameter in logging-while-drilling, can reflect rock composition, pore structure, and mechanical properties in real time, theoretically correlating with rock-breaking efficiency [10,11]. Drilling practices in the ultra-deep continental shale of the Fengcheng Formation in the Mahu Sag also indicate a significant correlation between resistivity and mechanical drilling rate. Therefore, a systematic investigation of resistivity response mechanisms in this region elucidates the linkage between rock-breaking efficiency and resistivity. This enables the development of resistivity logging-based models for predicting rock-breaking efficiency and dynamically optimizing drilling parameters. This is of significant importance for overcoming the over-reliance on field experience in current drilling parameter adjustments and advancing the development of ultra-deep continental shale oil.

Current research on the correlation mechanisms between shale rock-breaking efficiency and resistivity logging responses is still in its nascent stage. Existing studies indicate that rock-breaking efficiency is primarily related to rock mechanical properties and drilling engineering parameters [12–14], while shale resistivity is synergistically controlled by factors such as pyrite and clay mineral content, organic matter graphitization degree, porosity, and fluid saturation [15–19]. Cao et al. [20,21] employed multiple experimental methods, including scanning electron microscopy, TOC testing, X-ray diffraction whole-rock analysis, porosity, and resistivity testing, and found that shale resistivity is mainly influenced by pyrite and clay mineral content. Dong et al. [22] used petrophysical experiments and three-dimensional digital core simulations to demonstrate that shale resistivity is jointly affected by framework structure, pore fluids, and mineral composition, with clay minerals like montmorillonite and illite reducing resistivity through cation exchange. Li et al. [23] further observed that the layered structure and high surface area of clay minerals promote micropore development in shale reservoirs, thereby increasing bound water saturation and reducing resistivity. Additionally, Clegg et al. [24] conducted electromagnetic three-dimensional inversion on ultra-deep shale and found that increased overburden pressure reduces rock porosity and fluid volume, thereby increasing resistivity. Shi et al. [25] performed petrophysical experiments on shales with varying properties under different saturation states, analyzing the influence of parameters such as porosity, permeability, and saturation on shale resistivity, revealing that it is comprehensively affected by rock physicochemical parameters. Zhou et al. [26,27] established a relationship between organic matter graphitization degree and resistivity using Raman spectroscopy and petrophysical experiments, discovering that higher graphitization significantly reduces shale resistivity. Nie et al. [28] analyzed two-dimensional SEM-EDS images of shale reservoirs to characterize mineral distributions. They applied Markov chain-Monte Carlo (MCMC) methods to generate three-dimensional digital cores, demonstrating that shale resistivity is governed by clay mineral content, pyrite content, pore types, and organic matter graphitization degree. Cheng et al. [29] integrated mineral composition analysis, electron microscopy, total organic carbon (TOC) content, water saturation measurements, and logging data. Using random modeling techniques, they constructed three-dimensional numerical core models and simulated shale resistivity responses through finite element analysis. The results indicate that increases in clay mineral content, pyrite content, water saturation, and organic matter graphitization degree all lead to decreased resistivity in shale reservoirs. However, existing research predominantly focuses on marine shale reservoirs and shallow-to-medium burial depths, with few reports on the genesis of high resistivity in ultra-deep shale from the Mahu Sag. The ultra-deep shale in the Mahu Sag is continental, featuring complex geological conditions, strong rock heterogeneity, numerous resistivity influencing factors, and intricate coupling relationships among them. Therefore, there is an urgent need to refine the resistivity response mechanisms in ultra-deep continental shale, reveal the genesis of high resistivity, and provide a theoretical basis for further exploring the association mechanisms between ultra-deep continental shale rock-breaking efficiency and resistivity.

This study systematically characterizes the mineral composition of ultra-deep shale in the Mahu Sag using X-ray diffraction (XRD) and petrophysical methods. It further investigates the effects of mineral composition, porosity, water saturation, temperature, and cementation on shale resistivity. These findings enhance the comprehension of electrical responses in ultra-deep shale, elucidate the interrelationships among resistivity parameters, rock mechanical properties, and rock-breaking efficiency, and facilitate the development of resistivity-based real-time models for predicting rock-breaking efficiency. This is expected to provide scientific support for addressing challenges such as high costs, low accuracy, and high difficulty in predicting rock-breaking efficiency in ultra-deep continental shale, as well as the lack of theoretical basis for drilling parameter adjustments.

2. Materials and Methods

The rock samples used in this study are authentic downhole specimens from the Permian Fengcheng Formation in the Mahu Sag of the Junggar Basin (Figure 1), representing typical ultra-deep continental shale in China. The mineral composition is dominated by felsic, dolomitic, and mixed minerals [30]. The experimental samples were cored at depths of approximately 4600–5000 m, with lithology characterized as mixed shale and porosity ranging from 2.3% to 6.6%.

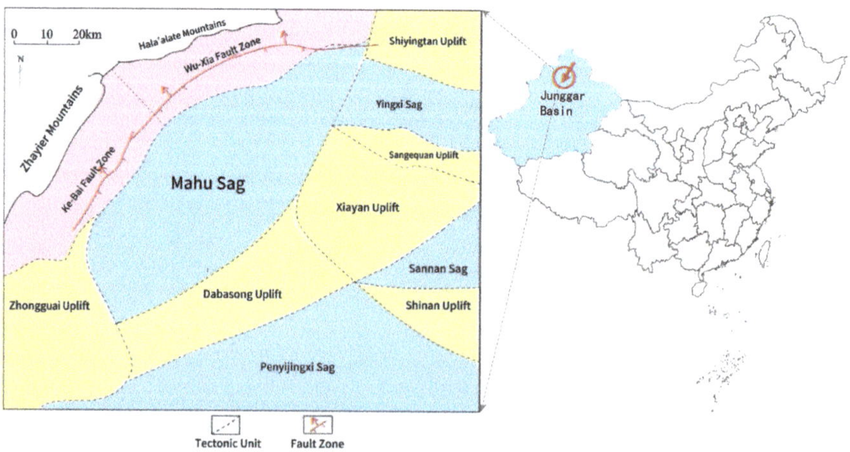

Figure 1. Location map of the Mahu Sag in the Junggar Basin, northwest China.

2.1. Sample Preparation

Cylindrical core plugs with a diameter of 25 mm and length of 50 mm were obtained from full-diameter cores using wire cutting. To minimize end-face effects on the plugs and ensure the parallelism of the upper and lower surfaces meets experimental requirements, the core plugs were mechanically polished using an MC004 grinder-polisher (LECO Corporation, St. Joseph, MI, USA). After polishing, the surface height variation ranged from 500 to 1000 nm, with parallelism of the upper and lower surfaces not exceeding 0.01 mm and vertical deviation not exceeding 0.05° (Figure 2).

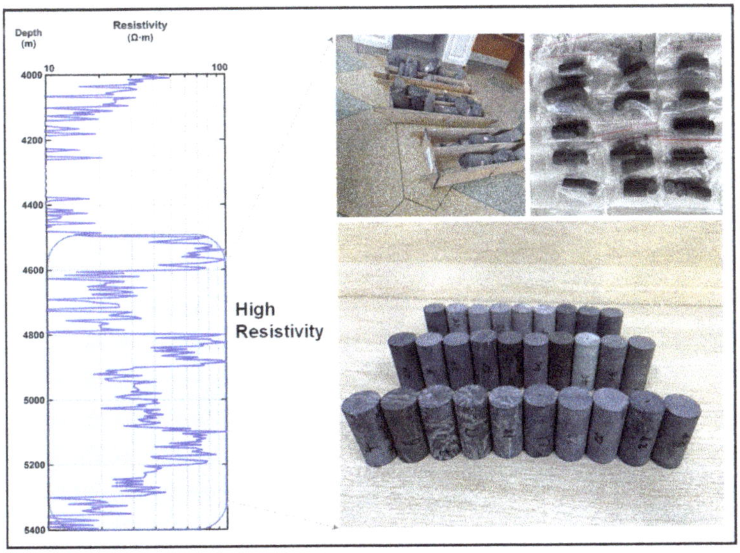

Figure 2. Selected high-resistivity rock samples based on logging resistivity.

2.2. Resistivity Testing

Empirical evidence in shale oil exploration indicates that resistivity thresholds are typically set at 100 Ω·m for high resistivity and 1000 Ω·m for ultra-high resistivity. Therefore, this study focuses on the ultra-high resistivity phenomenon in shale exceeding 1000 Ω·m. From a total of 29 rock sample groups, 12 cylindrical core plug samples with resistivity exceeding 1000 Ω·m were selected. All samples were dried in a muffle furnace at 100 °C for 24 h prior to testing to eliminate interference from pore water and free water. Resistivity measurements were conducted using an ST2643 ultra-high resistance tester (Beijing Oriental Jicheng Co., Ltd., Beijing, China) (measurement range: 1–10,000 Ω·m), strictly adhering to the Chinese national standard GB/T 1410-2006/IEC 60093:1980 [31] "Methods of Test for Volume Resistivity and Surface Resistivity of Solid Insulating Materials." The testing method employed the two-electrode approach, applying a direct current voltage to the upper and lower end faces of the sample and recording the current to calculate volume resistivity. To enhance measurement accuracy, the sample end faces were coated with metal electrodes prior to testing to minimize contact resistance effects. Other operations, such as drying and dimensional measurements, followed conventional laboratory procedures using standard equipment and are not detailed here.

2.3. XRD Analysis

The remaining core samples, after wire cutting, were crushed and ground into fine powders for X-ray diffraction (XRD) analysis. The powdered samples were evenly spread on a flat sample holder to minimize preferred orientation effects. Diffraction patterns were collected using a Bruker AXS D8 Advance A25 diffractometer (Karlsruhe, Germany) equipped with a Cu Kα radiation source operated at 40 kV and 40 mA. Scans were performed over a 2θ range of 2°–70°, with a step size of 0.02° and a scanning rate of 2° min^{-1}.

Quantitative phase analysis was performed using JADE 9.0 software (Materials Data Inc., Ashburn, VA, USA) integrated with the ICDD PDF-4+ 2019 database. The Reference

Intensity Ratio (RIR) method, also known as the K-value method, was employed, with α-Al_2O_3 (corundum) serving as the internal reference material. RIR (K) values for all identified phases were obtained from the PDF-4+ 2019 database. The weight fraction (W_X) of each phase was determined according to:

$$W_X = \frac{\frac{I_X}{K_X}}{\sum_i \frac{I_i}{K_i}}$$

I_X is the integrated intensity of the primary diffraction peak of phase X, K_X is its RIR value relative to corundum, and the denominator represents the sum over all detected crystalline phases. A fixed amount of corundum was added to each sample as an internal standard to correct for matrix effects, with background subtraction and peak fitting conducted in JADE 9.0.

Relative standard uncertainties in the calculated phase fractions were derived from triplicate analyses, accounting for peak overlap, background noise, and instrumental variability. These uncertainties ranged from 3% to 10%, corresponding to 3%–5% for major phases (>20 wt%), 5%–8% for minor phases (5–20 wt%), and 8%–10% for trace phases (<5 wt%). These results are consistent with established benchmarks for RIR-based quantitative XRD analysis, where the absolute uncertainty is typically less than $\pm 50\ X^{-0.5}$ wt% at the 95% confidence level, with X denoting the weight fraction.

3. Results

3.1. Mineral Characterization

Based on the X-ray diffraction (XRD) whole-rock quantitative analysis of 12 samples (Figures 3 and 4), the mineral composition of ultra-deep shale in the Mahu Sag exhibits the following characteristics. (1) Dominance of brittle minerals: Quartz and feldspar are predominant, with total contents typically ranging from 50% to 70%; quartz varies significantly between wells (Ma49 $\approx$ 60%; Xia207 $\approx$ 30%), while feldspar ranges mostly from 10% to 30% (average $\approx$ 20%), collectively imparting high brittleness to the reservoir. (2) Carbonates display a bimodal distribution, with dolomite and calcite occurring ubiquitously. Select samples exhibit pronounced enrichment (>30%; e.g., dolomite $\approx$ 40% in Xia207, calcite $\approx$ 32.9% in Xia203), whereas others contain <15% carbonates. This variability reflects alternating terrigenous clastic inputs and alkaline lacustrine chemical precipitation during sedimentary-diagenetic evolution. (3) Low clay content (sample range $\approx$ 3–15%, group average $\approx$ 6%), indicating transformation or compaction of primary argillaceous components due to deep burial diagenesis. (4) Accessory minerals dominated by pyrite (generally 2–4%, highest in Ma49 $\approx$ 3.4%), with local detection of hornblende (Ma5103 $\approx$ 0.5%), suggesting localized anoxic–reducing depositional environments and possible volcanic ash input. In summary, quartz, feldspar, and carbonates (collectively averaging $\approx$ 80%) dominate the mineral composition of ultra-deep shale in the Mahu Sag. This assemblage confers high brittleness and yields a pore system dominated by microfractures and intergranular pores. The XRD analytical patterns of the samples are provided in Appendix A.

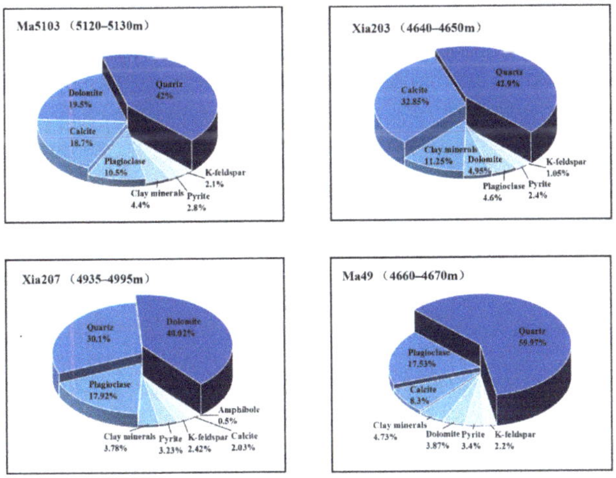

Figure 3. XRD mineral composition analysis for various well areas in the Mahu Sag.

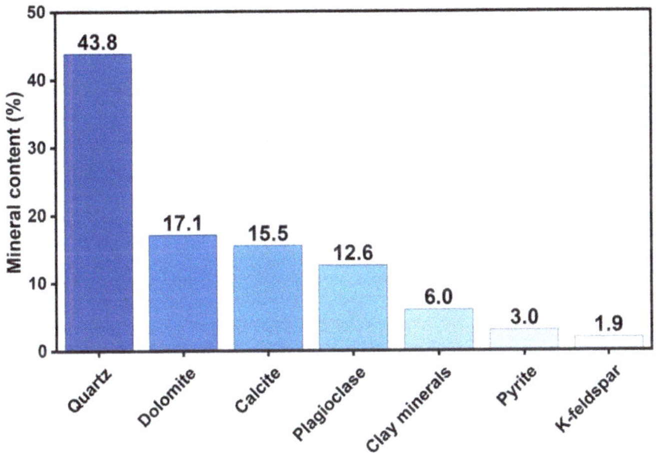

Figure 4. Average XRD mineral composition analysis in the Mahu Sag.

3.2. Analysis of Resistivity

3.2.1. Mineral Composition

Mineral composition, as the fundamental building block of the shale framework, directly determines charge carrier migration paths, thereby controlling apparent resistivity. Intrinsic resistivities of minerals differ markedly. Insulating minerals, such as quartz and dolomite, elevate rock resistivity due to their high intrinsic values, which occupy pore spaces and form dense frameworks. In contrast, conductive minerals like pyrite and certain feldspars reduce overall resistivity by creating low-resistance pathways through semiconductor behavior or ionic migration. Synergistic interactions among minerals further complicate electrical responses and may be influenced by thermal evolution, recrystallization, and microstructural heterogeneity, rendering traditional effective medium models (e.g., Bruggeman equation) not directly applicable. To systematically evaluate the comprehensive regulation of mineral composition on resistivity, this study utilized X-ray diffraction quantitative analysis results, combined with logarithmically transformed experimental resistivity measurements, to generate a correlation heatmap for multivariate visualization analysis.

The results (Figure 5) show that pyrite exhibits a significant negative correlation with resistivity (r = −0.59), indicating its strong inhibitory effect as a conductive mineral. The presence of pyrite also reflects deposition under an anoxic environment, which favored the preservation of organic matter, potentially influencing pore-filling and conductivity characteristics. Feldspar displays a moderate negative correlation (r = −0.32), suggesting its potential to introduce conductive paths via ion migration. Clay minerals and calcite show very weak correlations with resistivity (r = −0.04 and −0.09), indicating limited contributions to conductivity in dry samples. In contrast, dolomite and quartz exhibit weak positive correlations (r = 0.21 and 0.017), reflecting the potential contributions of their insulating intrinsic properties. Hornblende, due to its extremely low content (average 0.25%), has a near-zero correlation (r = −0.014). Furthermore, cross-correlations among minerals reveal synergistic and competitive relationships, such as the negative correlation between quartz and feldspar (r = −0.22), indicating mutually exclusive distribution of aluminosilicate minerals, and between dolomite and calcite (r = −0.49), reflecting competitive precipitation of carbonate minerals during diagenesis. Overall, the high resistivity of ultra-deep shale in the Mahu Sag is not dominated by a single mineral but results from the mutual balance between conductive and insulating minerals, shaped by both mineral composition and depositional environment.

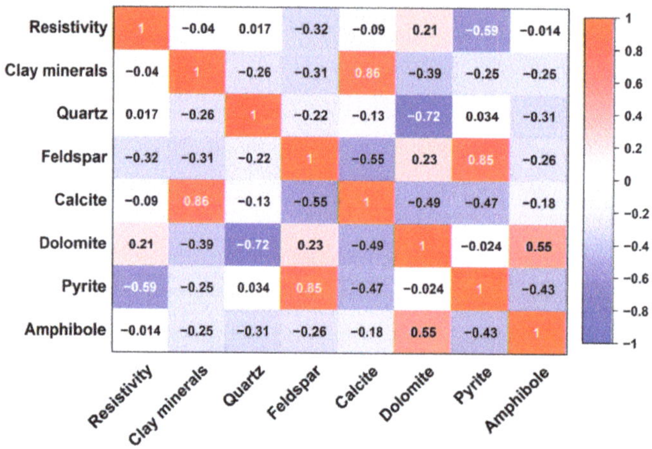

Figure 5. Pearson correlation heatmap of resistivity and mineral composition.

Strongly Correlated Minerals

Pyrite emerges as the sole strongly correlated mineral, with contents ranging from 1.9% to 4.8% (average 3.1%) and a moderate negative Pearson coefficient (r = −0.588; Figure 6), signifying its pivotal role in suppressing resistivity [32]. This statistical significance positions pyrite as a key antagonist in high-resistivity formation within the Mahu Sag ultra-deep shale. As a sulfide mineral, pyrite's intrinsic semiconductor properties, characterized by resistivity below 10^{-2} Ω·m, enable efficient electron conduction via its cubic lattice, fostering interconnected networks that bypass insulating matrices [27].

In the dry-sample regime—achieved through 100 °C drying to exclude fluid interference—pyrite's p-type semiconduction persists, with iron vacancies serving as hole acceptors, maintaining conductivity independent of electrolytes. This contrasts sharply with fluid-saturated models like Archie's, where electrolytic paths dominate; here, solid-state mechanisms prevail, allowing even low pyrite fractions to drastically lower resistivity by establishing preferential current routes. Regression analysis of the dataset quantifies this impact: a 1% pyrite increase correlates with a 15%–20% resistivity drop, often spanning orders of magnitude as thresholds for conductive bridging are met in heterogeneous fabrics. Geologically, the Mahu Sag's high thermal maturity (>110 °C) and depths (>4500 m) may induce pyrite alterations, such as partial oxidation or isotopic shifts, potentially narrowing its band gap and amplifying conduction under geostress.

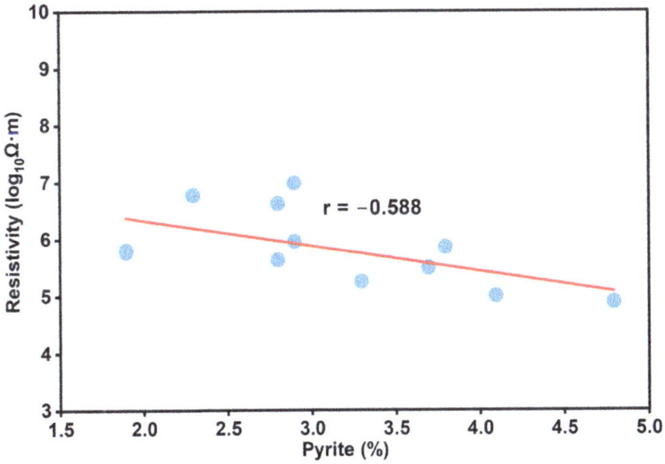

Figure 6. Correlation between resistivity and pyrite content in ultra-deep shale samples from the Mahu Sag (r = −0.588).

This strong correlation elucidates pyrite's disruptive influence on resistivity logs, where localized enrichments could mask broader insulating trends, complicating petrophysical interpretations. In conventional shale reservoirs, comparative studies frequently attribute low-resistivity zones to pyrite enrichment [33]. Ultimately, pyrite's influence exemplifies the mineral assemblage's complexity: its conductivity suppresses high resistivity while informing reservoir evaluation strategies. This highlights the necessity of integrating mineralogical and electrical datasets in ultra-deep continental settings.

Weakly Correlated Minerals

Feldspar and dolomite represent weakly correlated minerals, exerting transitional influences on resistivity in the Mahu Sag shale. Feldspar (potassium and sodium-calcium variants) contents vary from 5.1% to 33.8% (average 16.8%), displaying a moderate negative correlation (r = −0.319; Figure 7, left), suggesting a secondary suppressive effect. As framework aluminosilicates, feldspars harbor mobile cations (K^+, Na^+, Ca^{2+}) that facilitate ionic conduction through defects, boundaries, or adsorption, even in dry states [34]. In diagenesis under ultra-deep burial regime, processes like albitization and partial dissolution under elevated temperatures and pressures liberate ions, modestly enhancing local conductivity and reducing overall resistivity.

The correlation's weakness implies feldspar's subordinate status relative to pyrite's networks or carbonate insulation. Compositional variability—Na-plagioclase promoting greater ion mobility than K-feldspar—introduces heterogeneity, with samples exceeding 20% feldspar showing 10%–30% lower resistivities. This aligns with effective medium approximations, positioning feldspar as a semi-conductive diluent in insulating matrices. In dry conditions, where aqueous conduction is absent, the influence of ion-exchangeable clays diminishes compared to the hydrated conduction predicted by the Waxman–Smits model.

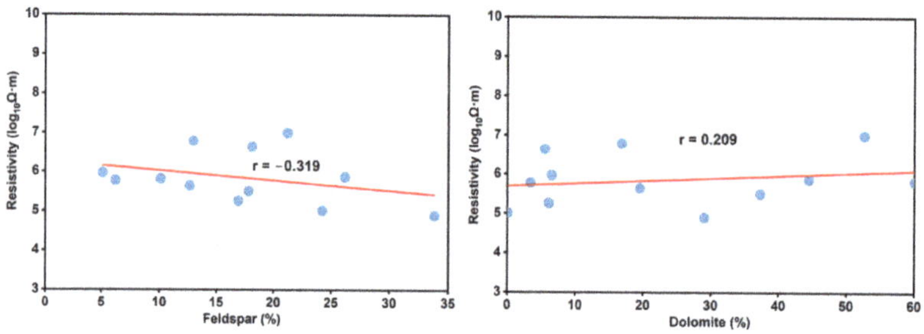

Figure 7. Correlations between resistivity and feldspar (**left**, r = −0.319) and dolomite (**right**, r = 0.209) contents in ultra-deep shale samples from the Mahu Sag.

Conversely, dolomite ranges from 0% to 60% (average 22.1%), with a weak positive correlation (r = 0.209; Figure 7, right), indicating mild promotional tendencies without significance. As a carbonate, dolomite's moderate resistivity (~10^6–10^8 Ω·m) and role as authigenic cement or infill densify the structure, impeding conductive routes by increasing tortuosity and sealing micropores [33]. In the alkaline lacustrine Mahu Sag, dolomite forms rhombohedral intergrowths with silicates during evaporation-driven precipitation, enhancing matrix insulation post-recrystallization under burial loads.

Regression reveals a 5%–10% resistivity rise per 10% dolomite increase, attributed to its thermal stability preventing decomposition, unlike some calcites. This positive trend counters conductive phases, with air-filled pores in dry samples amplifying isolation. Cross-correlations (e.g., r = −0.49 with calcite) reflect Mg-Ca competition in diagenesis, modulating availability. Studies on continental shales corroborates carbonates' insulating contributions in low-fluid settings, but here the weak link emphasizes interactive dilution by silicates or sulfides [27].

Together, feldspar's modest suppression and dolomite's subtle enhancement illustrate balanced weak correlations, highlighting multivariate controls in resistivity. For accurate modeling, decoupling these via finite-element simulations or core-scale experiments is crucial, informing ultra-deep reservoir strategies where such minerals bridge strong influencers and neutrals.

Essentially Uncorrelated Minerals

Clay minerals, quartz, calcite, and hornblende constitute essentially uncorrelated components, with minimal to negligible impacts on resistivity, reflecting their passive roles in the Mahu Sag ultra-deep shale. Clays range from 1.6% to 15.5% (average 5.7%), yielding a negligible correlation (r = −0.004; Figure 8a), indicating irrelevance in dry conditions. Conventionally, layered clays (illite, montmorillonite) lower resistivity via cation exchange and bound-water conduction in Waxman-Smits models, but drying at 100 °C eliminates interlayer hydration, rendering lattices insulating [33,35–37]. Ultra-deep burial drives smectite-illite transitions, dehydrating structures and curtailing exchange capacity, neutralizing electrical influence amid dominant frameworks.

Quartz, dominant at 16.8%–67.1% (average 39.8%), shows a near-zero positive correlation (r = 0.017; Figure 8b), lacking significance despite high intrinsic resistivity (>10^{12} Ω·m) and stability. As the primary brittle constituent, quartz theoretically enhances insulation owing to its volumetric dominance and the associated pore-space reduction caused by compaction and cementation [38]. However, synergies with feldspar (cross-r = −0.22) or clays dilute this, as enrichment often accompanies conductive aluminosilicates, positioning quartz as a foundational but non-driving element in resistivity increase.

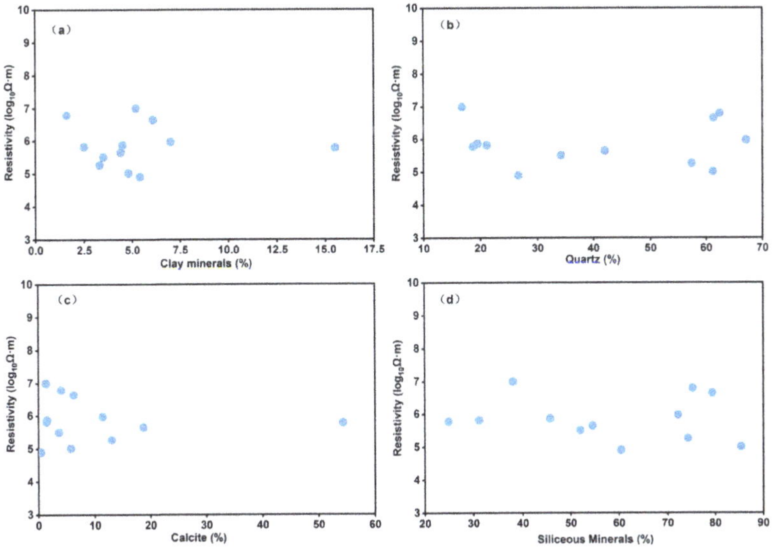

Figure 8. Correlations between resistivity and clay minerals (**a**), quartz (**b**), calcite (**c**), and siliceous minerals (**d**) contents in ultra-deep shale samples from the Mahu Sag.

Calcite parallels dolomite in carbonate function but exhibits a weak negative correlation (r = −0.09; Figure 8c), with negligible conductivity due to moderate resistivity and limited ion mobility (Ca^{2+}) in dry, compacted states. Diagenetic rivalry (cross-r = −0.49 with dolomite) constrains its precipitation, yielding minor variability without substantial suppression. Hornblende, a representative amphibole derived mainly from volcanic detritus, is scarce (0.25% on average) and shows no meaningful correlation (r = −0.014) with resistivity, indicating negligible conductive contribution.

Additionally, siliceous minerals, predominantly comprising quartz with minor contributions from other silica-rich phases, were examined for their correlation with resistivity in the ultra-deep Mahu Sag shale. Content varies from 16.8% to 85.4% (average 58.1%), reflecting variable detrital input and diagenetic silica mobilization. Pearson correlation analysis yields a coefficient of −0.117 (Figure 8d), indicating a negligible weak negative relationship that fails to achieve statistical significance ($p > 0.05$). This outcome underscores the limited insulating influence of siliceous minerals, despite their inherently high volume resistivity (>10^{12} $\Omega \cdot m$) and structural stability, which theoretically should enhance overall rock insulation through framework densification and pore occlusion. However, interactions with conductive or semi-conductive phases, such as feldspar (cross-correlation r = −0.22), dilute this potential, as siliceous enrichment often co-occurs with aluminosilicate variations that introduce minor ionic pathways. In dry conditions, the absence of fluid-mediated conduction further marginalizes their role, positioning siliceous minerals as passive components in high-resistivity rather than primary drivers. These findings suggest that in continental ultra-deep environments, siliceous content alone does not dictate electrical properties, highlighting the need for holistic multivariate models that capture interactions among multiple mineral constituents.

These uncorrelated minerals emphasize ensemble-driven resistivity, where diagenetic evolution in high-temperature (>110 °C), high-pressure (>80 MPa) settings minimizes fluid-reliant effects, rendering them electrically inert. Variance analysis attributes < 5% resistivity fluctuation to these, contrasting strong correlates like pyrite. Comparative shale studies affirm similar neutrality in low-clay, silicate-rich systems, advocating petrophysical models that prioritize interactions over isolates. This insight refines resistivity understanding, suggesting targeted exclusions in simulations for enhanced prediction in continental ultra-deep contexts.

3.2.2. Porosity

In the study of ultra-deep shale reservoirs, resistivity serves as a core geophysical parameter, commonly used to characterize reservoir fluid saturation and pore structure. Traditional models (e.g., the Archie equation) are primarily applicable to sandstone reservoirs under saturated fluid conditions, where resistivity typically exhibits a negative correlation with porosity, as conductive fluids like formation water filling the pores reduce overall resistivity. This study focuses on ultra-deep shale in the Mahu Sag, employing dried core samples for experiments to exclude fluid influences and concentrate on the intrinsic relationship between rock framework resistivity and porosity. This approach aids in revealing the contributions of mineral composition and microstructure to the genesis of high resistivity, potentially involving the distribution of insulating minerals (e.g., quartz or calcite) or thermal evolution effects.

Table 1 lists the porosity and resistivity measurement data for ultra-deep shale in the Mahu Sag under dry rock sample conditions. Porosity ranges from 0.023 to 0.066, with an average of 0.0437; resistivity ranges from 7.9×10^4 to 9.9×10^6 Ω·m, with an average of 2.0×10^6 Ω·m, reflecting significant high-resistivity characteristics consistent with the low conductivity of the dry rock framework.

Table 1. Resistivity and porosity test data for samples.

Sample No.	Resistivity (Ω·m)	Porosity
1	79,433	0.023
2	103,148	0.027
3	734,587	0.048
4	318,572	0.045
5	182,633	0.036
6	985,196	0.047
7	9,874,321	0.066
8	441,383	0.026
9	4,343,182	0.063
10	6,088,605	0.054
11	604,340	0.037
12	657,461	0.052

Pearson correlation analysis indicates a significant positive correlation between porosity and resistivity, with a correlation coefficient $r = 0.749$ (Figure 9). As porosity increases, resistivity tends to rise. Linear regression model fitting results show that the 95% confidence interval confirms reliable model parameters, with residual diagnostics revealing no significant heteroscedasticity or nonlinear deviations. In dried rock samples, conductive fluids (e.g., formation water) in the pore space are completely removed, and pores are primarily filled with high-resistivity air [27]. At this point, the rock's conductivity almost entirely depends on the ionic and electronic conductivity of the solid framework minerals themselves, as well as the limited contact points between mineral particles. Elevated temperatures induce thermal expansion of the rock framework, thereby enlarging pore spaces and increasing porosity. This enhances the volumetric fraction of highly insulating air within the rock, while proportionally reducing the conductive solid mineral matrix capable of current transmission. The presence of insulating pores directly replaces potential conductive paths while also altering the flow of current through the solid framework.

Elevated porosity reduces inter-particle contact points, thereby increasing contact resistance. This compels current to follow longer, more tortuous paths through the residual framework, diminishing the effective cross-sectional area for conduction and prolonging pathway lengths, which ultimately elevates apparent resistivity [39].

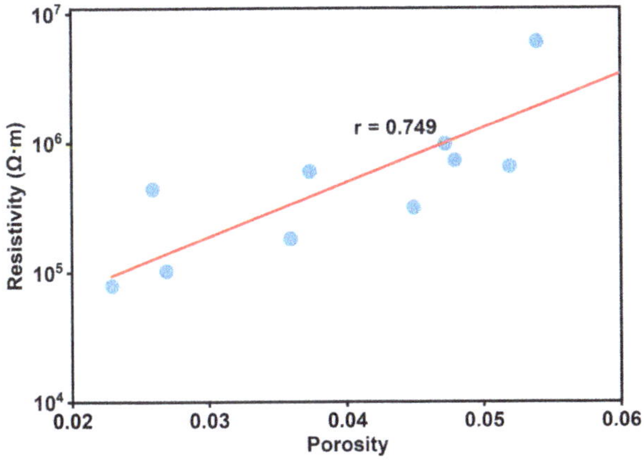

Figure 9. Correlation between resistivity and porosity.

3.2.3. Temperature

The resistivity measurements of twelve ultra-deep shale samples from the Fengcheng Formation in the Mahu Sag, Junggar Basin, China, were conducted at two distinct temperatures: 22.5 °C (ambient conditions) and 100 °C (simulating elevated reservoir thermal regimes). The results, presented in Figure 10, reveal a consistent inverse relationship between temperature and electrical resistivity across all samples. At 22.5 °C, the resistivity values ranged from 7.94×10^4 Ω·m (Sample No. 1) to 9.87×10^6 Ω·m (Sample No. 7), with a mean value of 2.03×10^6. In contrast, at 100 °C, the resistivity decreased markedly, ranging from 1.08×10^3 Ω·m (Sample No. 2) to 2.48×10^5 Ω·m (Sample No. 7), yielding a mean of 4.40×10^4. This temperature-induced reduction is visually evident in the semi-logarithmic scatter plot (Figure 1), where blue points (22.5 °C) cluster at higher resistivity magnitudes compared to red points (100 °C), indicating a systematic downward shift.

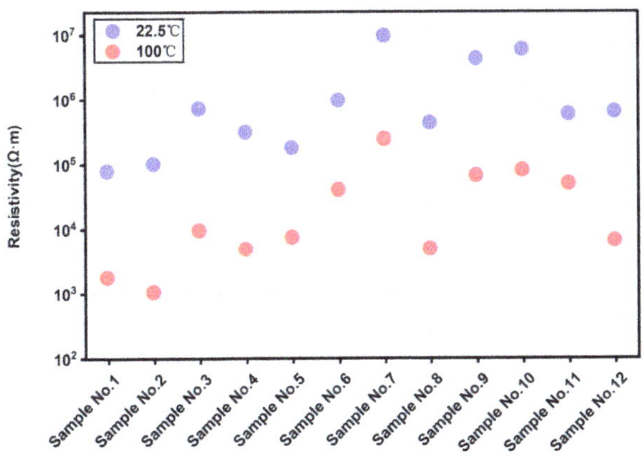

Figure 10. Resistivity at room temperature (22.5 °C) and high temperature (100 °C).

Statistical analysis further quantifies this correlation. The Pearson correlation coefficient between resistivity at the two temperatures is 0.9408 on a linear scale and 0.9176 on a logarithmic scale, underscoring a strong positive association despite the overall decline in values with increasing temperature. The average resistivity ratio (22.5 °C/100 °C) was calculated as 58.49, highlighting an approximately 58-fold decrease on average, though variability suggests influence from sample-specific mineralogical or microstructural factors.

To model the thermal dependence, an Arrhenius-type relationship was approximated for the resistivity behavior, assuming ionic conduction dominance in these shales: $\rho = \rho_0 \, e^{Ea/kT}$, where ρ is resistivity, Ea is the apparent activation energy, k is Boltzmann's constant, and T is absolute temperature. Using the two-point data, the computed Ea values ranged from 0.3033 to 0.5615 eV, with a mean of 0.4790. These energies align with thermally activated conduction processes typical in clay-rich sedimentary rocks [40].

The observed temperature-dependent resistivity decline in the Mahu Sag ultra-deep shales can be attributed to enhanced ionic mobility and solid-state conductivity at elevated temperatures, which are prevalent mechanisms in argillaceous formations enriched with clay minerals such as illite and smectite. In these dry samples, the high resistivity at ambient conditions likely stems from the insulating effects of organic matter, pyrite dissemination, and low-porosity matrices, characteristic of the alkaline lacustrine depositional environment in the Fengcheng Formation [41]. As temperature rises to 100 °C, thermal agitation facilitates ion dissociation in bound water within clay interlayers and mineral defects, reducing overall resistivity. This is particularly pronounced in ultra-deep shales where maturation processes, including graphitization of organic matter, may further modulate conductivity.

The strong correlation coefficients indicate that mineralogical composition exerts a controlling influence, with outliers (e.g., Sample No. 7 exhibiting the highest resistivity and ratio) potentially reflecting higher organic content or fracture sealing, which impedes conduction pathways at lower temperatures but becomes less effective under thermal stress. The calculated activation energies (mean ~0.48 eV) suggest a conduction regime dominated by surface-bound water and electrolyte migration rather than electronic semiconduction, consistent with observations in analogous shale systems where salinity and clay content amplify temperature sensitivity [33]. This thermal sensitivity has implications for the genesis of high resistivity in the Mahu Sag: it underscores that anomalous high-resistivity zones may arise from low thermal maturity or diagenetic alterations that restrict ion mobility, rather than solely from hydrocarbon saturation.

Variability in the resistivity ratios (standard deviation ~28) highlights heterogeneity, possibly linked to differential pyrite oxidation or clay transformation during burial, which could be exacerbated in ultra-deep settings exceeding 5000 m depth. Comparative studies in other basins, such as the Bakken Formation, report similar thresholds where resistivity drops sharply above ~74 °C due to kerogen maturation, reinforcing that temperature acts as a key modulator of electrical properties in organic-rich shales [42].

3.3. Engineering Practice Analysis

Field logging-while-drilling (LWD) data from pilot horizontal wells in the Mahu Sag provide critical insights into resistivity patterns within the Permian Fengcheng Formation, highlighting the interplay between electrical properties and drilling performance. In wells MaHW5102 and MaHW5103, resistivity logs exhibit a progressive escalation with depth. Upon penetrating the C8 layer, values rise from approximately 100 to 400 Ω·m, reflecting a transition into more compact, low-porosity zones influenced by diagenetic compaction and mineral cementation. This increase intensifies in the C9-10 layers, peaking at 400–800 Ω·m and occasionally exceeding 4000 Ω·m in ultra-high resistivity intervals (Figure 11). These anomalies correspond to regions dominated by insulating minerals such as quartz and dolomite, as identified in prior XRD analyses, where reduced conductive phases like pyrite minimize charge pathways. The elevated resistivity signals zones of heightened rock resistance, characterized by strong heterogeneity and high geostress (>80 MPa), which exacerbate bit wear and fracture propagation challenges during drilling. Such patterns align with the formation's alkaline lacustrine origin, where burial depths exceeding 4500 m promote densification, further amplifying electrical insulation and correlating with reduced permeability.

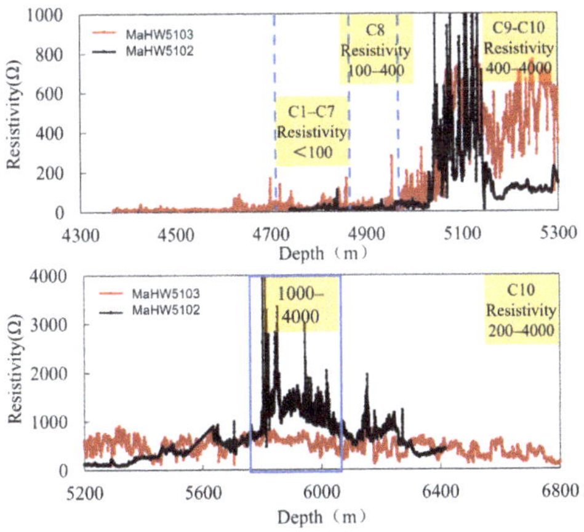

Figure 11. Resistivity logs of horizontal wells in the Mahu Sag pilot area.

To quantify the impact on drilling efficiency, cross-plot analysis was conducted between average rate of penetration (ROP) during trip drilling in the third spud section and average resistivity values across multiple wells. The data reveal a pronounced negative correlation, empirically modeled by the linear equation:

$$y = 25.615\, x^{-0.407}$$

with a coefficient of determination $R^2 = 0.78$ (Figure 12, left). This relationship indicates that as resistivity increases—driven by factors like low porosity and insulating mineral dominance—ROP declines exponentially on a logarithmic scale, confirming higher resistivity's association with harder, more brittle formations that demand greater mechanical energy for breakage. For instance, in intervals exceeding 1000 Ω·m, ROP often drops below 5 m/h, underscoring the bottleneck in ultra-deep shale operations where conventional parameters fail to adapt dynamically. This correlation integrates seamlessly with the resistivity elucidated earlier, where positive temperature coefficients and porosity effects further stiffen the rock matrix under in situ conditions.

The right panel of Figure 12 presents 3D resistivity models constructed via geostatistical interpolation of LWD data, visualizing spatial variations across the Fengcheng layers. The cross-sectional view depicts heterogeneous distributions, with high-resistivity cores (red hues > 1000 Ω·m) embedded in compact, low-conductivity zones, indicative of localized dolomite cementation and quartz enrichment. The layered model illustrates lateral and vertical gradients, revealing fault-influenced discontinuities that contribute to uneven stress fields and variable drillability. These models facilitate advanced reservoir characterization, enabling geosteering adjustments to avoid high-resistance barriers and target sweeter spots with optimal fracture potential.

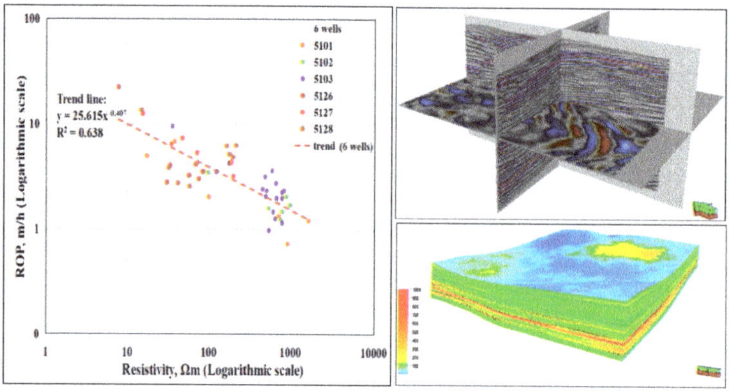

Figure 12. Cross-plot of average ROP versus average resistivity in the Mahu Sag (logarithmic scale) and 3D resistivity models of the Fengcheng Formation.

Collectively, these engineering analyses validate resistivity as a robust proxy for real-time drillability prediction in ultra-deep continental shale. By leveraging LWD resistivity for intelligent parameter optimization—such as weight-on-bit and rotary speed adjustments—drilling efficiency can be enhanced by up to 20%–30%, reducing operational costs and non-productive time. This approach not only addresses the "three highs and one strong" challenges of the Mahu Sag but also provides a scalable framework for global ultra-deep shale exploitation, bridging petrophysical insights with practical development strategies.

4. Discussion

Based on the XRD testing and petrophysical analysis results, the genesis of high resistivity in ultra-deep shale from the Mahu Sag can be summarized into the following four aspects: (1) Sedimentary environment and mineral composition. The Fengcheng Formation shale in the Mahu Sag formed in an alkaline lacustrine sedimentary environment, where terrigenous clastic and chemical deposits alternate. Under strong evaporation conditions, high-salinity lake water promoted the precipitation of carbonate minerals (e.g., dolomite) while inhibiting the massive formation of clay minerals; the provenance area provided abundant quartz and feldspar, with limited clay supply. This alkaline lacustrine sedimentary environment shaped the initial mineral composition characterized by high brittle minerals and low clay content in the shale, laying the mineral foundation for high resistivity. (2) Transformation by ultra-deep continental diagenesis. In the ultra-deep continental environment, the shale underwent intense compaction and cementation, resulting in a substantial reduction in primary pores and extremely low permeability. Meanwhile, clay minerals experienced dehydration and transformation (e.g., montmorillonite to illite) during deep burial, releasing interlayer water and reducing exchangeable cations, thereby decreasing rock conductivity. The high-temperature and high-pressure environment also prompted partial organic matter pyrolysis to generate bitumen or graphite, filling pores and expelling pore water, further reducing conductivity. Therefore, ultra-deep continental diagenesis renders the shale more compact, further elevating resistivity. (3) Pore fluid characteristics: The ultra-deep shale in the Mahu Sag typically has porosity less than 5%, dominated by nanoscale micropores and intercrystalline pores, with essentially no free water. The primary pore fillings are shale oil and minor bound water. Since bound water is adsorbed on mineral surfaces, ion migration is severely restricted, rendering its contribution to conductivity negligible. Thus, the shale's conductivity is almost entirely determined by the solid framework. In this state, the shale naturally exhibits extremely high resistivity. In contrast, conventional shallow shale often contains certain amounts of free water or brine, leading to lower resistivity. (4) Temperature-pressure conditions: The high temperature conditions in ultra-deep layers have a dual impact on resistivity: on one hand, high temperatures enhance ionic mobility in mineral lattices, increasing charge carrier activation and reducing resistivity; on the other hand, high temperatures may induce mineral phase transitions or decomposition, generating new conductive phases. However, samples from the study area indicate that at 100 °C, major mineral components remain stable without evident decomposition, with high temperature primarily facilitating the intrinsic conductivity characteristics of minerals in dry conditions. Under high-pressure conditions, rock pores decrease and fractures reduce, to a certain extent decreasing rock resistivity.

In summary, resistivity in Mahu Sag ultra-deep shale results from the coupled effects of sedimentary, diagenetic, mineralogical, fluid, and temperature-pressure controls. Mineral composition serves as the intrinsic controlling factor, while temperature and pressure further reinforce high-resistivity characteristics. Integrating resistivity analysis with logging data and mineralogical characterization enables enhanced prediction of mechanical properties, offering both scientific and practical value for drilling operations in ultra-deep continental shale.

5. Conclusions

Through comprehensive research on the mineral composition and resistivity of ultra-deep shale samples from the Mahu Sag, the following main conclusions are drawn:

(1) Mineral composition characteristics: The ultra-deep shale in the Mahu Sag is dominated by brittle minerals such as quartz and feldspar, along with carbonate minerals, with extremely low contents of clay and conductive minerals. The total content of quartz and feldspar is typically greater than 50%, carbonate minerals exhibit a wide distribution range, clay minerals average 5%, and minerals like pyrite are less than 5%. This mineral assemblage imparts high brittleness, high hardness, and high resistivity to the ultra-deep shale in the Mahu Sag.

(2) High dry rock sample resistivity: All dry rock sample resistivities fall within the range of 10^5 to 10^7 Ω·m, far exceeding the typical logging resistivity values of 10^2 to 10^4 Ω·m for conventional shale. Considering that logging resistivity represents the rock resistivity in the near-wellbore zone under in situ formation conditions, influenced by pore water. Some samples exhibit dry rock resistivities as high as 10^7 Ω·m; this anomalously high resistivity reflects the scarcity of conductive pathways in dry samples, with the high resistivity phenomenon primarily influenced by high-resistivity minerals.

(3) High-temperature (100 °C) resistivity tests indicate that the resistivity of Mahu shale significantly decreases with rising temperature, exhibiting a negative correlation with temperature. This contrasts with the behavior of conventional water-bearing rocks, suggesting that the conduction mechanism in dry samples is dominated by intrinsic mineral conduction. High temperatures enhance lattice vibrations, increase charge carrier mobility, and lead to decreased resistivity. This finding holds significant implications for understanding rock electrical properties in ultra-deep high-temperature environments.

(4) Mechanisms of high resistivity genesis: The high resistivity of ultra-deep shale in the Mahu Sag results from the combined effects of multiple factors, including sedimentation, diagenesis, minerals, fluids, and temperature-pressure conditions. (1) Regarding mineral composition, high quartz, feldspar, and carbonate contents form a high-resistivity framework, while low clay and pyrite levels minimize conductive pathways. Collectively, these mineral interactions drive the elevated resistivity of Mahu

Sag shale. (2) Ultra-deep burial results in extremely low porosity and minimal water content, rendering the contribution of pore fluids to conductivity negligible.

Author Contributions: Conceptualization, Y.Y.; methodology, D.Z.; validation, P.Z.; resources, S.T.; data curation, R.Z.; writing—original draft preparation, Y.Y.; writing—review and editing, D.Z.; supervision, Y.T.; funding acquisition, P.Z. and Z.M. All authors have read and agreed to the published version of the manuscript.

Funding: This research was funded by National Natural Science Foundation of China (NSFC) (Grant No. 52504047); Youth Science Fund Project of the Department of Science and Technology of Xinjiang Uygur Autonomous Region (2025D01B194); the Sub-project of the Key R&D Program of Xinjiang Uygur Autonomous Region (No. 2024B01017-3); and the Introduction Program for Young Doctors ("Tianchi Talents") in the Xinjiang Uygur Autonomous Region.

Data Availability Statement: Data are contained within the article.

Conflicts of Interest: Author Yangfei Yu was employed by the company Xinjiang Oilfield Company, PetroChina. The remaining authors declare that the research was conducted in the absence of any commercial or financial relationships that could be construed as a potential conflict of interest.

Appendix A

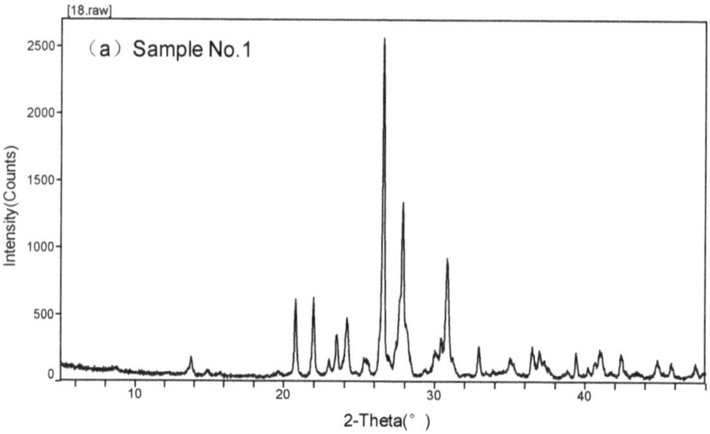

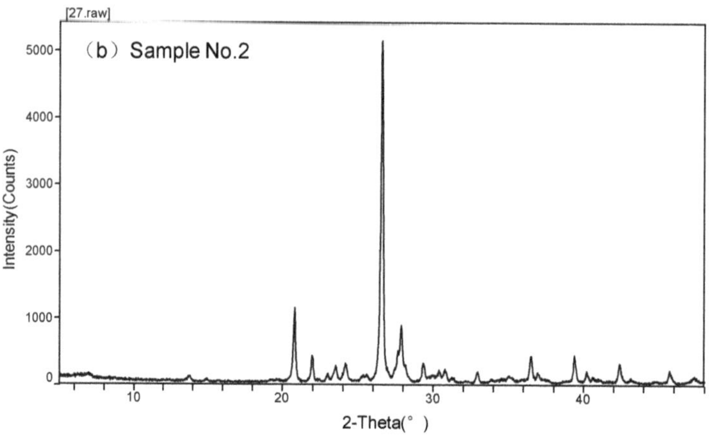

Figure A1. *Cont.*

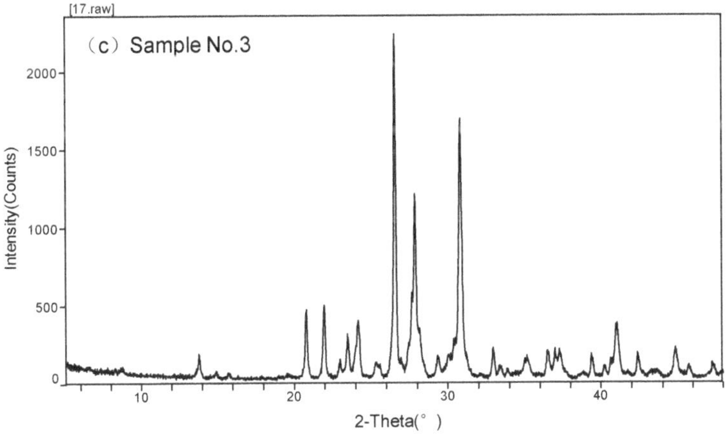

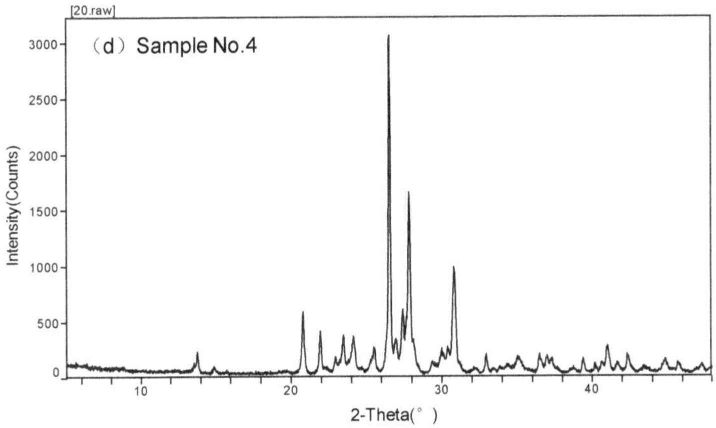

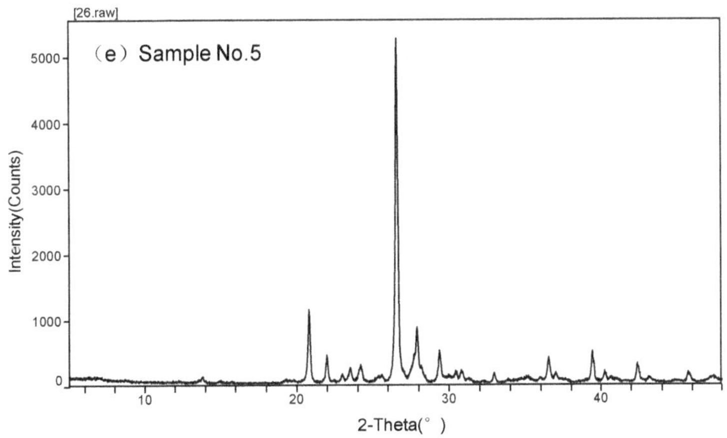

Figure A1. *Cont.*

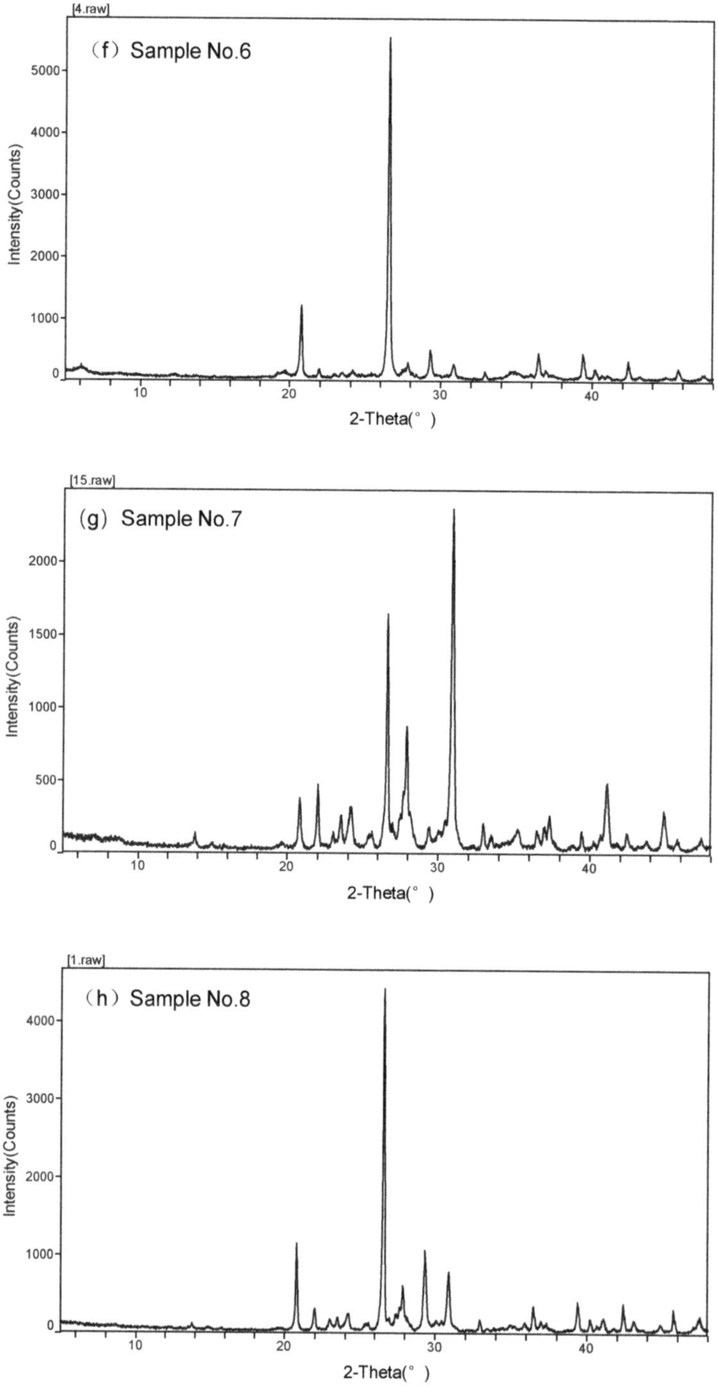

Figure A1. *Cont.*

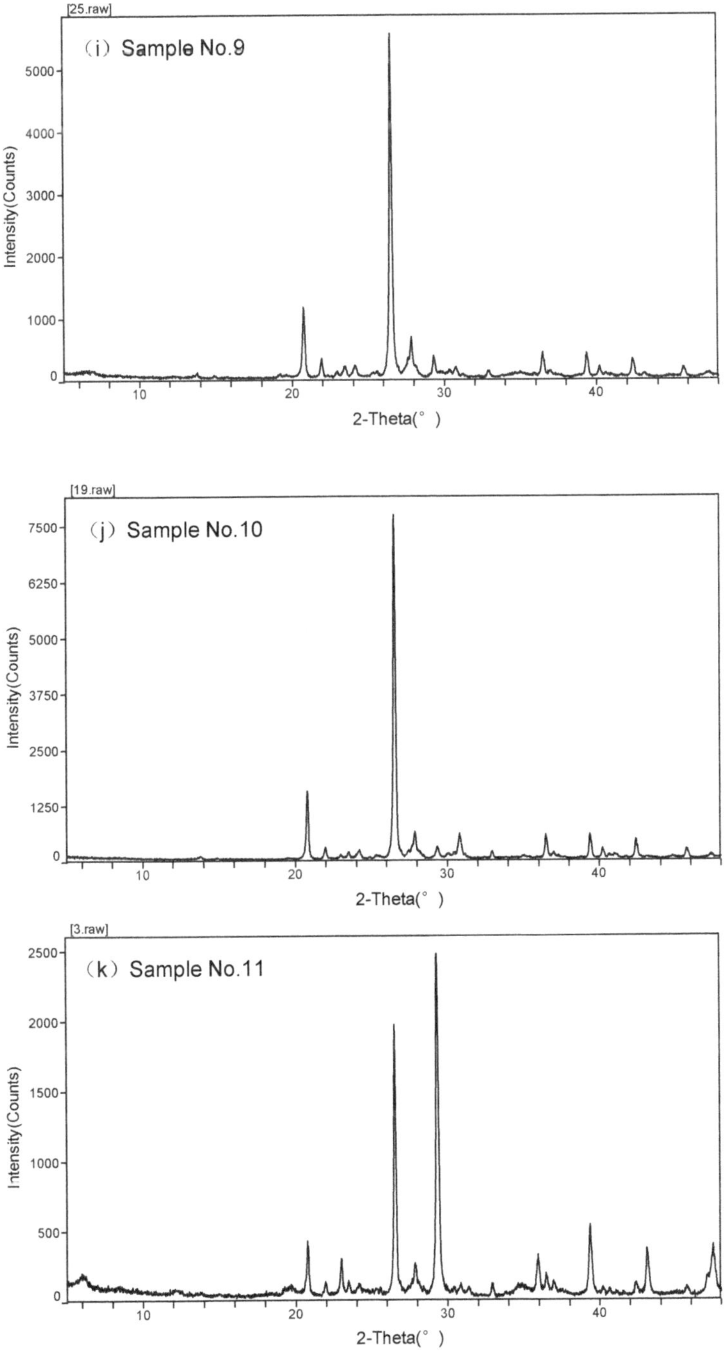

Figure A1. *Cont.*

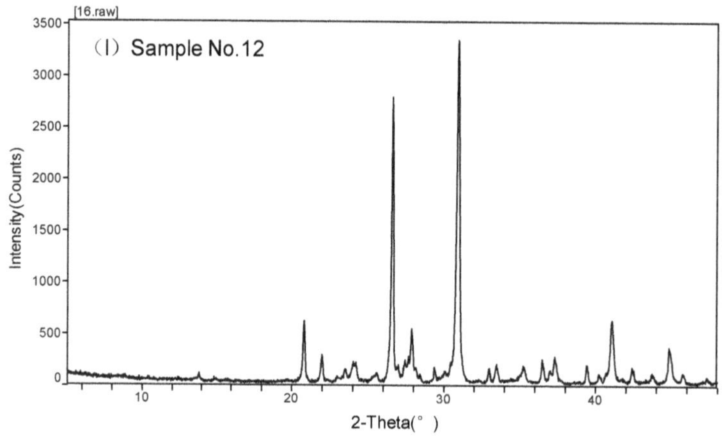

Figure A1. XRD patterns of 12 ultra-deep shale core samples (depths 4600–5000 m) from the Fengcheng Formation, Mahu Sag, China.

References

1. Guo, X.S.; Shen, B.J.; Li, M.W.; Liu, H.M.; Li, Z.M.; Zhang, S.C.; Yang, Y.; Guo, J.Y.; Liu, Y.L.; Li, P.; et al. Research progress and key research directions on shale oil in continental faulted lacustrine basins. *Pet. Explor. Dev.* **2025**, *52*, 1113–1127. [CrossRef]
2. Zhao, W.Z.; Zhu, R.K.; Zhang, J.Y.; Yang, J.R. Types of continental shale oil in China, exploration and development status and development trends. *China Pet. Explor.* **2023**, *28*, 1–13.
3. Jia, C.Z.; Zheng, M.; Zhang, Y.F. Unconventional hydrocarbon resources in China and the prospect of exploration and development. *Pet. Explor. Dev.* **2012**, *39*, 129–136. [CrossRef]
4. Zhang, D.Q.; Zhang, J.Q.; Wang, Y.F.; Tang, Y.; Yu, W.J. Progress and prospects of unconventional oil and gas exploration and development in China. *Resour. Sci.* **2015**, *37*, 1068–1075.
5. Huang, F.X.; Wang, S.Y.; Li, M.P.; Ouyang, J.L.; Liu, C.; Liu, H.Y.; Zeng, F.D.; Fan, J.J.; Jia, P. Progress and implications of deep and ultra-deep oil and gas exploration in China. *Nat. Gas Ind.* **2024**, *44*, 86–96.
6. Zhu, C.H.; Li, W.L. Unconventional oil and gas in China become an important support for increasing reserves and production. *China Pet. Enterp.* **2024**, *3*, 36–38.
7. Sun, Z.G.; Yu, C.L.; Chen, H.; Zhang, M.; Sun, Q.; Jia, L.H.; Sun, C.; Chen, T.; Zhang, H.X.; Fan, F. Current status and prospects of experimental technology for continental shale oil development. *Oil Gas Geol. Recovery Effic.* **2024**, *31*, 186–198.
8. Guo, X.S.; Li, M.W.; Zhao, M.Y. Development and utilization of shale oil and its role in energy. *Bull. Chin. Acad. Sci.* **2023**, *38*, 38–47.
9. Zhao, J.Z.; Yong, R.; Hu, D.F.; She, C.Y.; Fu, Y.Q.; Wu, J.F.; Jiang, T.X.; Ren, L.; Zhou, B.; Lin, R. Deep and ultra-deep shale gas fracturing in China: Problems, challenges and development directions. *Acta Pet. Sin.* **2024**, *45*, 295–311.
10. Zhang, X.Y.; Wang, J.F.; Guo, Y.J. Progress and development trends of logging while drilling technology. *Well Log. Technol.* **2006**, *1*, 10–15.
11. Yuan, H.Z.; Sun, J.M. Summary of new progress in logging technology. *Prog. Geophys.* **2005**, *3*, 786–795.
12. Dong, G.; Chen, P. A review of the evaluation, control, and application technologies for drill string vibrations and shocks in oil and gas well. *Shock Vib.* **2016**, *2016*, 7418635. [CrossRef]
13. Wang, D.Y.; Li, G.S.; Shi, H.Z.; Huang, Z.W.; Wu, Z.B. Progress and application of new methods for efficient rock breaking. *China Pet. Mach.* **2012**, *40*, 1–6.
14. Liu, Q.S.; Liu, J.P.; Shi, K.; Pan, Y.C.; Huang, X.; Liu, X.W.; Wei, L. Influence of rock brittleness indices on hob rock-breaking efficiency. *Chin. J. Rock Mech. Eng.* **2016**, *35*, 498–510.
15. Cheng, L.Z.; Wang, Y.; Zhang, C. Study on the anisotropy of complex resistivity in eastern Guizhou shale and its relationship with shale gas reservoir parameters. *Chin. J. Geophys.* **2021**, *64*, 3344–3357.
16. Sun, J.M.; Xiong, Z.; Luo, H.; Zhang, H.P.; Zhu, J.J. Analysis of low-resistivity genesis and logging evaluation of shale gas reservoirs in the Lower Paleozoic of the Yangtze region. *J. China Univ. Pet.* **2018**, *42*, 47–56.
17. Hao, J.Q.; Feng, R.; Zhou, J.G.; Qian, S.Q.; Gao, J.T. Discussion on the mechanism of resistivity changes during rock fracture process. *Chin. J. Geophys.* **2002**, *45*, 426–434.

18. He, G.S. Characteristics and genesis of low resistivity in marine shale in southeastern Sichuan. *Oil Gas Geol. Recovery Effic.* **2023**, *30*, 66–76.
19. Gao, H.Q.; Ding, A.X.; Cai, X.; He, G.S. Analysis of abnormal resistivity genesis in marine shale in the Middle-Upper Yangtze. *Fault-Block Oil Gas Field* **2016**, *23*, 578–582.
20. Cao, G.Q.; Zhang, B.; Yang, K.; He, X.L. Study on the differences in electrical characteristics based on shale brittleness: Taking shale from Wujiaping Formation in eastern Sichuan as an example. *Geophys. Prospect. Pet.* **2024**, *63*, 1075–1086.
21. Sun, F.; Sun, J.; Zeng, X.; Yuan, W.; Zhang, J.; Yan, W.; Yan, W. Analysis of the influencing factors on electrical properties and evaluation of gas saturation in marine shales: A case study of the Wufeng-Longmaxi formation in Sichuan Basin. *Front. Earth Sci.* **2022**, *10*, 824352. [CrossRef]
22. Dong, H.M.; Zeng, X.; Zhou, D.L.; Zhu, J.; Golsanami, N.; Sun, J.; Zhang, Y. Insights into the multiscale conductivity mechanism of marine shales from Wufeng-Longmaxi Formation in the southern Sichuan Basin of China. *J. Energy Eng.* **2023**, *149*, 04023008. [CrossRef]
23. Li, M.; Zhang, C. An improved method for improving the calculation accuracy of marine low-resistance shale reservoir parameters. *Front. Earth Sci.* **2022**, *10*, 1001287. [CrossRef]
24. Clegg, N.; Sinha, S.; Rodriguez, K.R.; Walmsely, A.; Sviland-Østre, S.; Lien, T.; Mouatt, J.; Marchant, D.; Schwarzbach, C. Ultra-deep 3D electromagnetic inversion for anisotropy, a guide to understanding complex fluid boundaries in a turbidite reservoir. In Proceedings of the SPWLA 63rd Annual Logging Symposium, Stavanger, Norway, 11–15 June 2022; p. D031S002R005.
25. Shi, Y.J.; Zhao, J.B.; Xiao, Z.S.; Xie, W.B.; Wang, J.B.; Zhang, X.Y.; Ke, S.Z.; Bai, S. Analysis of shale electrical dispersion characteristics and influencing laws based on complex resistivity experimental measurements. *Prog. Geophys.* **2025**, *40*, 1096–1104.
26. Zhou, T.; Cai, J.; Mou, S.; Zhao, Q.; Shi, Z.; Sun, S.; Guo, W.; Gao, J.; Cheng, F.; Wang, H.; et al. Influence of low-temperature hydrothermal events and basement fault system on low-resistivity shale reservoirs: A case study from the Upper Ordovician to Lower Silurian in the Sichuan Basin, SW China. *Minerals* **2023**, *13*, 720. [CrossRef]
27. Ma, X.; Wang, H.; Zhou, T.; Zhao, Q.; Shi, Z.; Sun, S.; Cheng, F. Geological controlling factors of low resistivity shale and their implications on reservoir quality: A case study in the southern Sichuan Basin, China. *Energies* **2022**, *15*, 5801. [CrossRef]
28. Nie, X.; Zou, C.C.; Meng, X.H.; Jia, S.; Wang, Y. Three-dimensional digital core modeling of shale gas reservoir rocks: Taking conductivity model as an example. *Nat. Gas Geosci.* **2016**, *27*, 706–715.
29. Cheng, J.; Yan, J.P.; Song, D.J.; Liao, M.J.; Guo, W.; Ding, M.H.; Luo, G.D.; Liu, Y.M. Low resistivity response characteristics and main controlling factors of shale gas reservoirs in the Ordovician Wufeng Formation-Silurian Longmaxi Formation in Changning area, southern Sichuan. *Lithol. Reserv.* **2024**, *36*, 31–39.
30. Jin, Z.J.; Zhang, Q.; Zhu, R.K.; Dong, L.; Fu, J.H.; Liu, H.M.; Yun, L.; Liu, G.Y.; Li, M.W.; Zhao, X.Z.; et al. Classification of continental shale oil in China and its significance. *Oil Gas Geol.* **2023**, *44*, 801–819.
31. *GB/T 1410-2006/IEC 62631-3-1:2016*; Methods of Test for Volume Resistivity and Surface Resistivity of Solid Insulating Materials. IEC: Geneva, Switzerland, 2006.
32. Zhou, Q.; Liu, J.; Ma, B.; Li, C.; Xiao, Y.; Chen, G.; Lyu, C. Pyrite characteristics in lacustrine shale and implications for organic matter enrichment and shale oil: A case study from the Triassic Yanchang Formation in the Ordos Basin, NW China. *ACS Omega* **2024**, *9*, 16519–16535. [CrossRef]
33. Liang, M.; Fu, G.; Yu, Y.; Yuan, H. Controlling factors of low resistivity in deep shale and their implications on adsorbed gas content: A case study in the Luzhou area. *Sci. Rep.* **2024**, *14*, 31861. [CrossRef]
34. Liang, F.; Zhao, Q.; Zhao, S.; Shi, Z.; Zhang, Q.; Zhou, T.; Wang, Y.; Cui, H.; Chen, H.; Wang, Y. Genesis mechanism of low resistivity in the Lower Cambrian Qiongzhusi Formation shale and its response characteristics to pore structure—Take the Z201 Well as an example. *ACS Omega* **2025**, *10*, 43995–44011. [CrossRef]
35. Li, Q.; Berraud-Pache, R.; Yang, Y.; Jaber, M. Biocomposites based on bentonite and lecithin: An experimental approach supported by molecular dynamics. *Appl. Clay Sci.* **2023**, *231*, 106751. [CrossRef]
36. Li, Q.; Zhu, D.Y.; Zhuang, G.Z.; Li, X.L. Advanced development of chemical inhibitors in water-based drilling fluids to improve shale stability: A review. *Pet. Sci.* **2025**, *22*, 1977–1996. [CrossRef]
37. Li, Q.; Berraud-Pache, R.; Souprayen, C.; Jaber, M. Intercalation of lecithin into bentonite: pH dependence and intercalation mechanism. *Appl. Clay Sci.* **2023**, *244*, 107079. [CrossRef]
38. Tan, X.; Wang, Z.; Jiao, P.; Wen, Z. Quantifying the pore characteristics and heterogeneity of the Lower Cambrian black shale in the deep-water region, South China. *ACS Omega* **2025**, *10*, 709–726. [CrossRef]
39. Zhu, L.; Ma, Y.; Cai, J.; Zhang, C.; Wu, S.; Zhou, X. Key factors of marine shale conductivity in southern China—Part II: The influence of pore system and the development direction of shale gas saturation models. *J. Pet. Sci. Eng.* **2021**, *207*, 109516. [CrossRef]
40. Rey, J.; Jongmans, D. Temperature dependence of the electrical resistivity of water-bearing rocks. *Geophysics* **1989**, *54*, 197–203.

41. Nazarenko, M.Y.; Kondrasheva, N.K.; Saltykova, S.N. Electrical resistivity of coal and oil shales. *Coke Chem.* **2018**, *61*, 345–349. [CrossRef]
42. Han, T.; He, H.; Fu, L.Y. New insights into how temperature affects the electrical conductivity of clay-free porous rocks. *Geophys. J. Int.* **2024**, *238*, 313–320. [CrossRef]

minerals

MDPI

Article

Influence of Hydration on Shale Reservoirs: A Case Study of Gulong Shale Oil

Feifei Fang [1], Ke Xu [2], Yu Zhang [3], Yu Wang [1,*], Zhimin Xu [4], Sijie He [5,*], Hui Huang [6], Hailong Wang [4], Weixiang Jin [1] and Yue Gong [1]

1 School of Petroleum Engineering, Chongqing University of Science and Technology, Chongqing 401331, China; fangfeifei@cqust.edu.cn (F.F.); jinweixiang@cqust.edu.cn (W.J.); gongyue@cqust.edu.cn (Y.G.)
2 Research Institute of Petroleum Exploration and Development, PetroChina, Beijing 100083, China; xk0929@163.com
3 Construction Project Management Branch, National Oil and Gas Pipeline Network Group Co., Ltd., Langfang 065000, China; zhangyu01@pipechina.com.cn
4 Research Institute of Exploration and Development, PetroChina Jilin Oilfield Company, Songyuan 138000, China; xuzhimin-jl@petrochina.com (Z.X.); wanghl-jlyt@petrochina.com.cn (H.W.)
5 School of Energy Resources, China University of Geosciences (Beijing), Beijing 100083, China
6 Fuyu Oil Production Plant, Jilin Oilfield Company, Songyuan 138000, China; huangh-jl@petrochina.con.cn
* Correspondence: 2022201043@cqust.edu.cn (Y.W.); hesijie@email.cugb.edu.cn (S.H.)

Abstract

In the process of the exploration and development of shale oil, the influence of hydration on shale reservoirs is complex, as it can not only improve porosity and permeability, but also lead to reservoir instability. At present, there is a lack of systematic understanding of the influence of hydration on the physical and chemical properties of shale oil reservoirs. Therefore, in this study, taking the Gulong shale oil reservoir in Songliao Basin as the research object, X-ray diffraction mineral composition analysis, electron microscope scanning, and micro-CT scanning were used to study the micro–macro-changes in shale caused by hydration, and the effects of different fracturing fluids on hydration were evaluated. The results show the following: (1) Hydration increases the porosity and permeability of Gulong shale through clay dispersion and dissolution pore formation, though these transient effects may compromise long-term reservoir stability due to pore-throat clogging. (2) Prolonged hydration significantly enhanced pore structure complexity, with tortuosity increasing by 64.7% (from 2.19 to 3.60) and the fractal dimension rising by 7.5% (from 1.99 to 2.14) with hydration time, and the proportion of larger pores (50–100 μm) increased significantly. (3) Hydration leads to crack propagation and new cracks, and the intersection of cracks reduces the core strength, which may eventually lead to macroscopic damage. (4) The influence of different fracturing fluids on the hydration reaction is obviously different. The higher the concentration, the stronger the hydration effect. Distilled water helps to increase porosity and permeability, but long-term effects may affect reservoir stability. The results of this paper reveal the changes in micro- and macro-characteristics of shale oil reservoirs under hydration, which is of great significance for analyzing the mechanism of hydration and provides theoretical support for improving shale oil recovery.

Keywords: hydration; shale oil reservoir; mineral composition; pore structure; fracturing fluid

https://doi.org/10.3390/min15080878

1. Introduction

With the gradual deepening of oil exploration and development of research into deep and complex strata, the stability of shale reservoirs has become increasingly prominent. Shale reservoirs often contain a large number of clay minerals, especially water-sensitive minerals such as montmorillonite and illite [1–3]. These minerals will undergo physical and chemical reactions such as expansion and dispersion under hydration, thus affecting the pore structure and physical and mechanical properties of the reservoir [4–6]. In the development of shale reservoirs, hydration may lead to changes in pore structure, which in turn affects the seepage capacity and strength of the reservoir [7,8]. In an environment with a high water content, hydration will cause the expansion of clay minerals, resulting in significant changes in porosity and permeability, especially affecting the stability and strength of the reservoir [9,10]. Therefore, the study of the effects of hydration on shale reservoirs not only helps to understand the physical and chemical properties of shale, but also provides a theoretical basis for further improving drilling efficiency and enhancing reservoir reconstruction.

With the extension of hydration time, the pore structure of shale may become more complex, and the porosity and permeability may increase, but this increase does not mean that the stability of the reservoir is improved [11,12]. On the contrary, hydration may lead to a decrease in reservoir strength, which in turn affects the safety and effectiveness of drilling and fracturing operations [13–15]. Therefore, it is of great theoretical value and practical significance to explore the specific influence mechanism of hydration on shale reservoirs and reveal its influence on reservoirs' physical properties and mechanical behavior. Gulong shale is a typical continental shale with high clay mineral content and a complex pore structure, which represent the characteristics of typical shale oil reservoirs in China [16–18]. Although some progress has been made in the exploration and development of Gulong shale in recent years, due to the influence of hydration, the problems of shale gas wellbore instability and unsatisfactory fracturing effects are still unexplored [6,19]. Therefore, studying the mechanism of hydration of Gulong shale and revealing its influence on reservoirs' physical properties and seepage characteristics is an important issue to be solved urgently in the field of shale gas development.

In this study, Gulong shale was selected as a representative shale reservoir for investigation. The effects of hydration on the physical and chemical properties and pore structure of the Gulong shale reservoir are systematically studied by means of X-ray diffraction analysis, micro-CT scanning, Electron microscope scanning analysis, and pulse porosity–permeability measurement experiment. Through a series of analytical characterization techniques, the effects of hydration on clay mineral expansion, porosity and permeability changes, and fracture propagation are discussed, and the evolution process of reservoir pore structure under different hydration times and different hydration strengths is studied.

2. Experimental Samples and Method

2.1. Experimental Samples

The samples used in this paper (Figure 1) were selected from Gulong shale in the first section of the Qingshankou Formation in the Songliao Basin. The total organic carbon (TOC) and X-ray diffraction analysis results are shown in Table 1. The TOC content of the three samples ranges from 2.21% to 2.33%. The brittle minerals of Gulong shale are mainly quartz and plagioclase, with quartz content of 30.2%–35.7% and plagioclase content of 11.4%–17.5%. The high content of brittle minerals provides favorable conditions for the formation of cracks, thus weakening the pressure-bearing capacity of the formation, which is one of the potential factors influencing wellbore instability. In addition, the content of clay minerals is as high as 40%, comprising mainly illite (32%–44%), followed by an illite/smectite mixed layer (25%–35%), which contains a small amount of chlorite (21%–43%).

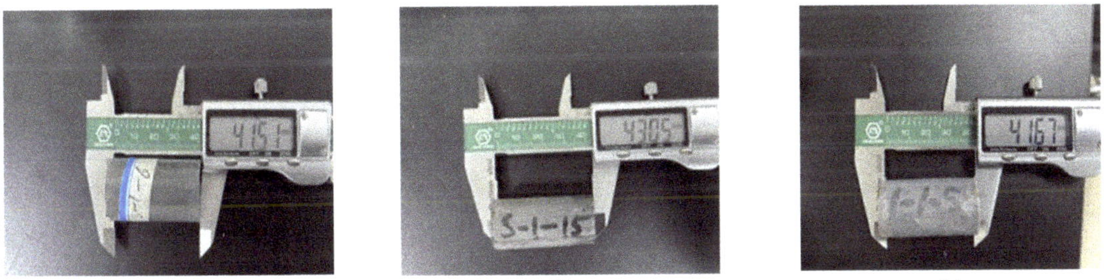

(a) (b) (c)

Figure 1. Rock core photos. (**a**) S-1-6; (**b**) S-1-15; (**c**) S-1-1.

Table 1. Integrated analysis of mineralogical composition and TOC content in rock samples.

Number	TOC	Mineral Content (%)					Relative Content of Clay Mineral (%)		
		Quartz	Potash Feldspar	Plagioclase	Pyrite	Clay Mineral	Mixed-Layer Illite	Illite	Chlorite
S-1-6	2.33	30.2	0.6	11.4	7.3	50.5	35	44	21
S-1-15	2.24	33.1	0.8	15.3	8.2	42.6	25	32	43
S-1-1	2.21	35.7	0.6	17.5	4.9	41.3	30	42	28

2.2. Experimental Instruments and Principles

2.2.1. X-Ray Diffraction Test

The Japanese Rigaku high-resolution SmartLab X-ray diffractometer (Japanese Rigaku, Tokyo, Japan) was used to perform X-ray diffraction analysis on the samples, analyze their diffraction patterns, and obtain information such as the composition of the material and the structure or morphology of the atoms or molecules inside the material. The diffraction characteristics of different crystal structures can be superimposed to determine the diffraction patterns of multiphase samples. By comparing the observed patterns with those in the powder diffraction database (PDF), the phase can be determined and the relative content of each phase can be estimated according to the diffraction intensity. X-ray diffraction analysis is conducted according to the procedures described in previously published studies [20,21].

2.2.2. Micron CT Scanning Test

A Domestic micro–nano dual-ray tube core CT scanning system is used to penetrate the core sample through X-ray, and the three-dimensional pore structure of the core is reconstructed through rotation and multi-angle projection images. This technology does not need to destroy the samples, and can quickly and non-destructively display the microscopic pore throat characteristics of reservoir rocks, and quantitatively analyze their geometry, size, connectivity, and distribution. The CT image is based on X-ray attenuation information, and the gray value is related to the core density, which can directly reflect the pore structure. A micron CT scanning test is conducted according to the procedures described in previously published studies [22]. Instrument Parameters: (1) Sample Parameters—1–260 mm (sample diameter). (2) Voltage—micron tube: 0–240 kV; nanotube: 0–180 kV. (3) Pixel Size—micron tube: 20–122 μm; nanotube: 0.6–20 μm.

2.2.3. Scanning Electron Microscopy Test

The electron microscope scanning analysis uses a finely focused electron beam that is stationary or performs grating scanning on the surface of the sample to bombard the surface of the sample so that it produces various signals (secondary electrons, backscattered electrons, Auger electrons, characteristic X-rays, photons of different energies, etc.), and the solid material is analyzed by using an electromagnetic lens system. The scanning electron microscopy test is conducted according to the procedures described in previously published studies [23,24]. The experimental parameters are as follows: sample dimensions—flake-shaped samples with diameters less than 25 mm; pixel size—0.9 nm–800 nm; accelerating voltage—500 V–30 kV; beam current—1 pA–100 nA; cutting precision—10 nm.

2.3. Experimental Method

2.3.1. X-Ray Diffraction Analysis

Shale is primarily composed of matrix pores and bedding fractures, and is rich in a variety of clay minerals. Chlorite has lipophilicity, its crystal structure is stable, and it is less affected by hydration. Kaolinite exhibits low hydrophilicity and negligible swelling capacity, whereas illite and montmorillonite represent the most expansive phyllosilicates upon hydration. Therefore, in this experiment, Gulong shale, with a high content of illite and illite–montmorillonite mixed-layer minerals was selected. In order to analyze the influence of different fracturing fluids on clay minerals in shale reservoirs, an X-ray diffractometer was used to test and analyze the content of clay mineral components. (1) Three Gulong shale cores were selected, each of which was divided into four slices, and the test samples were ground to about 0. 15 mm. (2) The shale samples were immersed in distilled water and different types of fracturing fluid systems (No.1 friction reducer: polyacrylamide-based system; No.2 friction reducer: guar gum-based system) with varying concentrations (0.1%, 0.3%) in a constant-temperature chamber. After soaking for 1, 3, 5, and 7 days (simulated field frac fluid residence time), the shale slices were carefully wiped dry and subsequently subjected to X-ray diffraction analysis. Permeability and porosity measurements were conducted before and after the treatment.

2.3.2. Micron CT Scanning Analysis

(1) A piece of Gulong shale was selected and vacuumized with a vacuum pump for 12 h, and then immersed in a beaker filled with distilled water. (2) After soaking for 0 days, 1 day, and 7 days, the samples were taken out, dried, and placed in the micro–nano dual-ray tube core CT scanning system for CT scanning analysis.

2.3.3. Scanning Electron Microscopy Analysis

(1) Three Gulong shale core slices were selected, and the surface of the slices was polished by argon ion polishing and sprayed with carbon film. (2) The treated shale sections were placed in a resistance-reducing agent solution containing distilled water and different types of resistance-reducing agent (No. 1 resistance-reducing agent is a polyacrylamide system, and No. 2 resistance-reducing agent is a guar gum system). After soaking in the incubator for 0 days and 7 days, the shale sections were wiped clean and subjected to electron microscope scanning experiments.

3. Result

3.1. Effect of Hydration on Microstructure and Physical Parameters of Shale

3.1.1. The Differences in Porosity and Permeability Before and After Hydration

The test specimens were submerged in distilled water and maintained at a standard temperature (25 °C) and pressure (1 atm) for a 7-day hydration period. The porosity and permeability of three Gulong shale cores were measured by a pulsed porosity and permeability instrument, following the procedures described in previously published studies [25,26]. The porosity before hydration was between 0.833% and 1.80%, and the porosity after hydration was between 0.915% and 2.37% (Figure 2a). The permeability before hydration was between 0.001 mD and 0.005 mD, and the permeability after hydration was between 0.0015 mD and 0.0076 mD (Figure 2b). Hydration will increase shale porosity and permeability, and improve the seepage channel of the reservoir to a certain extent. Xue et al. [9] also supported that the porosity and permeability parameters of rock samples increased after hydration, and the cracks that formed after hydration contributed greatly to the increase in the permeability of the rock samples.

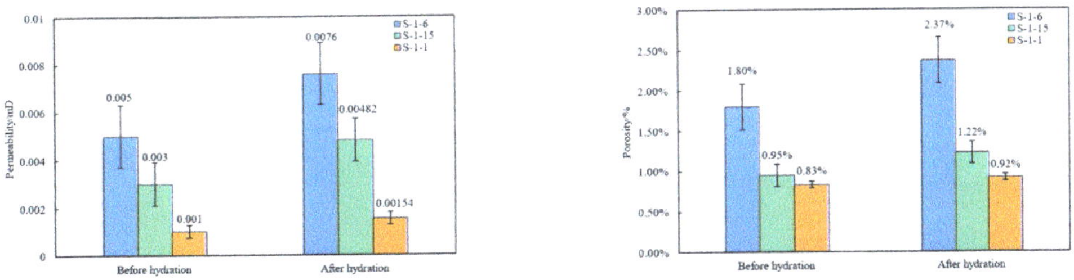

(a) (b)

Figure 2. Differences in porosity and permeability of Gulong shale before and after hydration. (**a**) Permeability change; (**b**) Porosity change.

3.1.2. Comparison of Shale Hydration Based on Micron CT Scan

The pore structure of shale reservoirs is complex, so the fractal dimension is often used to evaluate the complexity and heterogeneity of shale reservoirs [27,28]. The large fractal dimension of shale indicates that the complexity of its pore structure is high, and the roughness of its pore surface and the irregularity of its shape are high. Tortuosity, an important parameter affecting the permeability of porous media, characterizes the bending degree of pores with different sizes in porous media [29,30]. Figure 3 shows the pore identification model of the S-1-6 core before and after hydration. From Figure 1, it can be seen that the pore volume after hydration increases significantly, and the porosity of the S-1-6 core decreases first and then increases; the initial clay content is 1.87%, and it is reduced to 1.25% after 1 day, indicating that the early clay minerals expand and disintegrate after encountering water, which will fill up the small pores and cause the pore volume to decrease. The pore size increased significantly after 7 days of hydration, indicating that over time, the hydration of hydrophilic mineral particles increased and the cementation strength decreased, resulting in the loosening and shedding of mineral particles and the generation of dissolution pores, which eventually increased the pore volume. Figure 4 is a model of a shale stick before and after hydration. From Figure 4, it can be seen that the flow path of fluid in shale becomes more tortuous, and the tortuosity increases from 2.186 to 3.60. Figure 4 also shows that with the strengthening of hydration, a small number of dissolution pores are produced in shale, and these tiny dissolution pores make the pore structure of shale more complex. Figure 5 illustrates shale segmentation before and after hydration. It can be seen from Figure 5 that with an increase in hydration time, crack propagation and derivation occur on the surface of shale, and the fractal dimension increases (from 1.99 to 2.14). Figure 5 also shows that after hydration, the micropores and cracks in shale increase, and the fractal dimension increases, which makes the pore structure of shale more complex. These cracks intersect and connect with each other, reducing the strength of the core and causing macroscopic damage to the rock. Figure 6 shows the pore size distribution after hydration for different times. It can be seen from Figure 6 that the proportion of medium and large pores in shale increases gradually, and the growth rate of pores with sizes in the range of 100–150 μm is the largest, indicating that with an increase in hydration time, the formation of dissolution pores contributes to an increase in the proportion of macropores within the pore system. Previous studies by Yang [31] and Shi et al. [32] have demonstrated, through micro-CT and field emission scanning electron microscopy (FE-SEM) analyses, that water–rock interactions can induce the development of new micropores and microfractures in shale formations. This also shows that under the action of hydration with distilled water, Gulong shale will undergo structural damage, resulting in a decrease in the mechanical strength of the reservoir, and with the extension of soaking time, the greater the degree of damage, the greater the decrease in the rock's mechanical strength.

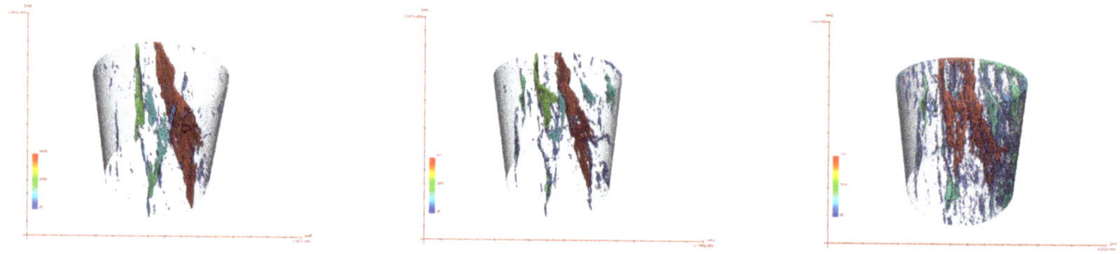

(a) (b) (c)

Figure 3. Hydration pore identification model. (**a**) Soaking for 0 days; (**b**) Soaking for 1 day; (**c**) Soaking for 7 days.

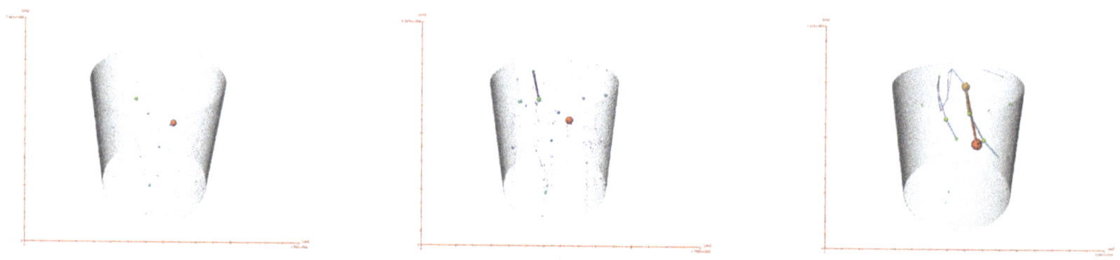

(a) (b) (c)

Figure 4. Shale ball stick model. (**a**) Soaking for 0 days; (**b**) Soaking for 1 day; (**c**) Soaking for 7 days.

(a) (b) (c)

Figure 5. Shale CT 3D scanning images. (**a**) Soaking for 0 days; (**b**) Soaking for 1 day; (**c**) Soaking for 7 days.

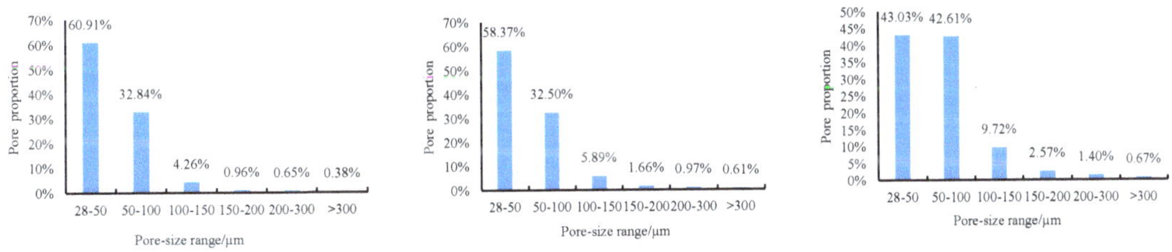

(a) (b) (c)

Figure 6. Shale pore size distribution histogram. (**a**) Soaking for 0 days; (**b**) Soaking for 1 day; (**c**) Soaking for 7 days.

3.2. Study on the Influence of Fracturing Fluid Type on Shale Hydration

3.2.1. Effect of Fracturing Fluid Type on Permeability

Table 2 shows the differences in porosity and permeability before and after hydration under different types of fracturing fluid conditions. It can be seen from the table that the fracturing fluid causes a decrease in permeability and porosity due to retention and adsorption of the rock pore throat surface. With an increase in concentration, the greater the decrease, the more serious the damage. At the same concentration, the damage degree of the guar gum system is greater than that of polyacrylamide system, and distilled water will increase the porosity and permeability of the reservoir.

Table 2. Changes in porosity and permeability before and after hydration of different fracturing fluids.

Experimental Scheme	Permeability (Before Hydration)/mD	Permeability (After Hydration)/mD	Rate of Change/%	Porosity (Before Hydration)/%	Porosity (After Hydration)/%	Rate of Change/%
1-A	0.005	0.0076	55.12	1.80	2.17	20.56
1-B	0.0051	0.00422	−20.85	1.81	1.71	−5.81
1-C	0.0055	0.004175	−31.74	1.82	1.55	−17.65
1-D	0.0053	0.004164	−27.28	1.80	1.58	−13.64
2-A	0.003	0.00482	60.58	0.95	1.14	20.32
2-B	0.0032	0.00268	−19.40	0.96	0.86	−11.11
2-C	0.0033	0.002405	−37.21	0.95	0.82	−16.65
2-D	0.0038	0.002944	−29.08	0.97	0.85	−12.64
3-A	0.001	0.00154	54.32	0.83	0.97	15.85
3-B	0.0011	0.00089	−23.60	0.82	0.73	−12.11
3-C	0.0013	0.000955	−36.13	0.82	0.69	−18.65
3-D	0.0012	0.000956	−25.52	0.81	0.71	−14.64

Scheme coding definitions: sample number-soaking solution (1: S-1-6; 2: S-1-15; 3: S-1-1)—A: distilled water; B: 0.1% No. 1 drag-reducing agent (polyacrylamide system); C: 0.3% No. 1 drag-reducing agent (polyacrylamide system); D: 0.1% No. 2 drag-reducing agent (guar gum system).

3.2.2. Comparison of Shale Hydration Based on X-Ray Diffraction

Tables 3 and 4 illustrate the differences in clay mineral content before and after shale hydration under different types of fracturing fluid conditions. As shown in Tables 3 and 4, there is little difference in the types of clay minerals before and after hydration with different types of fracturing fluids. The influence of different fracturing fluids on the reservoir is mainly due to the retention of micro-pore throats and adsorbed rock pore throat surfaces.

Table 3. Clay mineral content after soaking for 0 days.

Number	Experimental Scheme	Relative Content of Clay Minerals (%)			Ratio of Mixed Layer (%S)
		Mixed-Layer Illite	Illite	Chlorite	Mixed-Layer Illite
1	1-0-A	34	41	25	10
2	1-0-B	33	40	24	10
3	1-0-C	32	39	23	10
4	1-0-D	33	40	22	10
5	2-0-A	24	33	43	10
6	2-0-B	24	31	45	10
7	2-0-C	23	34	43	10
8	2-0-D	23	32	45	10
9	3-0-A	33	45	22	10
10	3-0-B	30	42	28	10
11	3-0-C	28	43	29	10
12	3-0-D	32	43	25	10

Scheme coding definitions: sample number-soaking solution (1:S-1-6; 2:S-1-15; 3:S-1-1)—A: distilled water; B; 0.1% No. 1 drag-reducing agent (polyacrylamide system); C: 0.3% No. 1 drag-reducing agent (polyacrylamide system); D: 0.1% No. 2 drag-reducing agent (guar gum system).

Table 4. Clay mineral content after soaking for 7 days.

Number	Experimental Scheme	Relative Content of Clay Minerals (%)			Ratio of Mixed Layer (%S)
		Mixed-Layer Illite	Illite	Chlorite	Mixed-Layer Illite
1	1-0-A	35	40	25	10
2	1-0-B	35	39	26	10
3	1-0-C	38	37	25	10
4	1-0-D	36	41	23	10
5	2-0-A	26	31	43	10
6	2-0-B	27	29	44	10
7	2-0-C	26	30	44	10
8	2-0-D	29	29	42	10
9	3-0-A	35	42	23	10
10	3-0-B	32	43	25	10
11	3-0-C	29	42	29	10
12	3-0-D	30	41	29	10

Scheme coding definitions: sample number-soaking solution—A: distilled water; B: 0.1% No. 1 drag-reducing agent (polyacrylamide system); C: 0. 3% No. 1 drag-reducing agent (polyacrylamide system); D: 0.1% No. 2 drag-reducing agent (guar gum system).

3.2.3. Comparison of Shale Hydration Based on Electron Microscopy Scanning Experiments

By comparing the shale slices of S-1-6 shale before and after hydration with different fracturing fluids (Figure 7), it was found that the shale samples contained more clay minerals before being soaked in distilled water. In addition, the edges of mineral particles were obvious, and there were large intergranular pores in the form of slits. When the shale sample was soaked for 7 days, it was found that the mineral particles fell off and dissolved, the tiny pores were further dissolved, and the pore radius increased. After soaking in the polyacrylamide system fracturing fluid for 7 days (Figure 8), it was found that the clay minerals had expanded significantly, and the edges of the particles were passivated. There was loosening and shedding of the particles, and there may have been clogging of the pores. After S-1-6 shale was soaked in guar gum fracturing fluid for 7 days (Figure 9), it was found that the clay minerals expanded significantly, and there was a phenomenon of fracturing fluid retention. A layer of hydration film was added on the surface of the clay minerals, and the number of tiny pores was reduced. In summary, hydration significantly affects the seepage capacity and stability of the reservoir by changing the pore structure and physical and mechanical properties of the shale. In the process of shale oil and gas development, reasonable formulas and construction parameters for fracturing fluid should be selected according to the hydration characteristics, so as to minimize the negative impact of hydration on reservoir stability and provide a scientific basis for shale reservoir development [33,34].

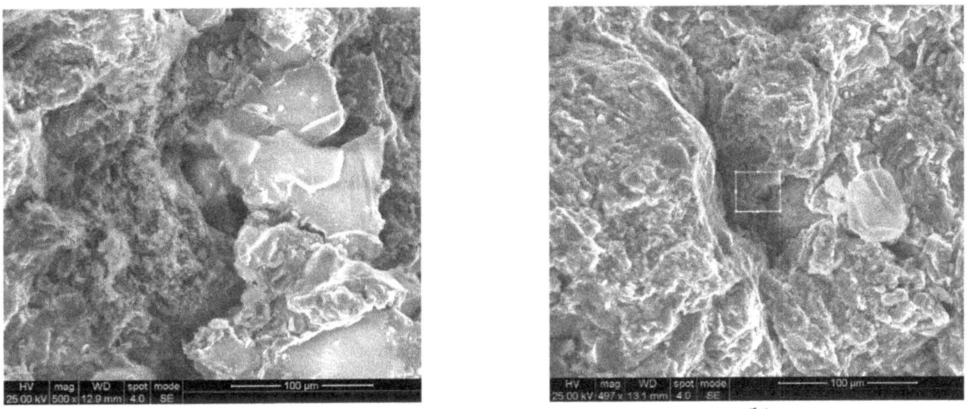

(a) (b)

Figure 7. Sample S-1-6 soaked in distilled water. (**a**) Before soaking; (**b**) After soaking.

(a) (b)

Figure 8. Sample S-1-6 soaked in 0.1% polyacrylamide fracturing fluid. (**a**) Before soaking; (**b**) After soaking.

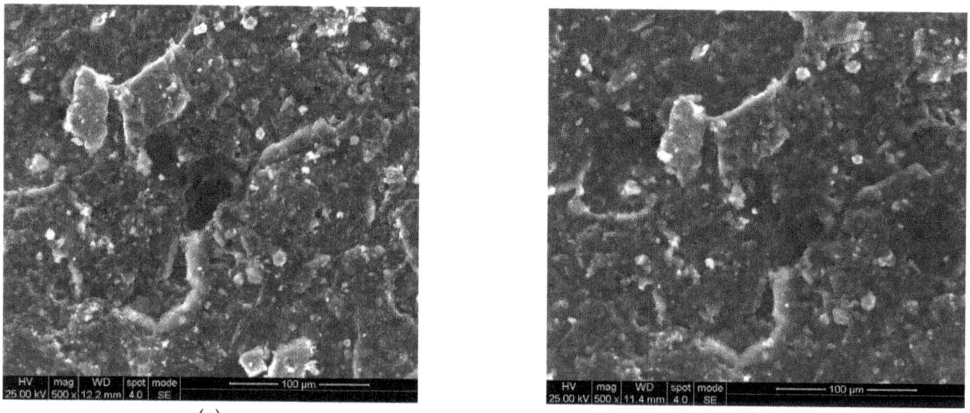

(a) (b)

Figure 9. Sample S-1-6 soaked in 0.1% guar gum fracturing fluid. (**a**) Before soaking; (**b**) After soaking.

4. Discussion

In this study, the changes in reservoirs' physical properties under hydration and influence mechanism of hydration on Gulong shale in the Songliao Basin were systematically evaluated by means of CT scanning, X diffraction, micron CT scanning and electron microscope analysis. The results show that hydration has a significant effect on the pore structure and permeability of shale reservoirs. The expansion effect of clay minerals and the properties of fracturing fluid components are the main reasons for the weakening of pore connectivity and the change in permeability. In addition, the three-dimensional visualization results of micro-CT scanning further reveal the dynamic evolution process of reservoir pore structure complexity in response to hydration, which provides a deeper understanding the mechanism of changes in shale reservoir microstructure. In this study, the influence of fracturing fluid type on the hydration of the Gulong shale oil reservoir was quantitatively and qualitatively analyzed by X-ray diffraction mineral composition analysis and the electron microscope scanning method for the first time, which provided an important reference for reservoir evaluation and development optimization.

Although this study reveals the important influence of hydration on reservoir reconstruction, it is found that the hydration of shale for a long enough time will produce more positive effects, such as micro-fracture propagation and dissolution pore generation. However, there are still some differences between the experimental conditions and the actual formation environment. In field operation, it is also possible that the confining pressure in the high-pressure environment inhibits the cracking and further expansion of the original micro-cracks, resulting in the closure of micro-cracks rather than their expansion. The complex ion environment in the actual formation may have a more complex effect on the expansion behavior of minerals and the change in pore structure. Therefore, the influence of shale reservoir hydration in a real environment needs further exploration. In addition, due to the limited number of samples, the regional applicability of our research results needs to be further verified. Therefore, future research should be carried out on the following topics: first, the regulatory effects of different chemical environments on hydration should be explored by introducing experiments that simulate complex formation conditions; second, researchers should expand the scope of research samples and evaluate the universality of hydration under different geological backgrounds; thirdly, the mechanisms of different chemical additives in fracturing fluid should be further studied to provide more scientific guidance for reservoir reconstruction.

5. Conclusions

(1) The expansion and dispersion of clay minerals and the generation of dissolution pores caused by hydration are the main reasons for the increase in the porosity and permeability of the reservoir, but also reduce its stability.

(2) The analysis of the pore identification model shows that hydration enhances the complexity and heterogeneity of shale pore structure. Hydration makes the fluid flow path more tortuous. The tortuosity and fractal dimension increase with an increase in hydration time. The tortuosity increases from 2.186 to 3.60, and the fractal dimension increases from 1.99 to 2.14. Dissolution causes an increase in the proportion of medium and large pores, especially in the range of 100–150 μm.

(3) Hydration induces the propagation and derivation of cracks on the surface of shale, and the intersection of cracks reduces the strength of the core, which leads to macroscopic damage. Scanning electron microscopy showed that the hydration of clay minerals made the edges of mineral particles loosen and fall off, resulting in dissolution, thus forming larger intergranular pores.

(4) The fracturing fluid of the guar gum system causes more serious damage to the reservoir than that of the polyacrylamide system, and the higher the concentration, the greater the decrease in porosity and permeability. Distilled water promotes a relative increase in porosity and permeability, but its long-term effect may lead to instability of the reservoir structure.

Author Contributions: Conceptualization, S.H., F.F. and K.X.; methodology, S.H. and Y.Z.; software, Y.Z.; validation, Y.W. and Z.X.; formal analysis, H.H.; investigation, S.H. and F.F.; resources, H.W. and W.J.; data curation, Y.G.; writing—original draft preparation, Y.W., S.H. and F.F.; writing—review and editing, K.X., S.H. and F.F.; visualization, S.H.; supervision, F.F. and W.J.; project administration, H.H.; funding acquisition, S.H. and F.F. All authors have read and agreed to the published version of the manuscript.

Funding: This research was funded by Research on the key technologies for chemical flooding to enhance oil recovery in low-permeability/tight reservoirs (2023ZZ17YJ04), Study on hydration damage mechanism in shale reservoirs (202X-KFKT-30) and General Program of Chongqing Natural Science Foundation (CSTB2025NSCQ-GPX0934).

Data Availability Statement: The data presented in this study are available in this article.

Conflicts of Interest: Author K.X. was employed by the Research Institute of Petroleum Exploration and Development, PetroChina. Author Y.Z. was employed by the Construction Project Management Branch, National Oil and Gas Pipeline Network Group Co., Ltd. Author Z.X. and H.W. was employed by the Research Institute of Exploration and Development, PetroChina Jilin Oilfield Company. Author H.H. was employed by the Fuyu Oil Production Plant, Jilin Oilfield Company. The remaining authors declare that the research was conducted in the absence of any commercial or financial relationships that could be construed as a potential conflict of interest.

References

1. Zhang, D.; Meegoda, J.; Goncalves da Silva, B.; Hu, L. Impact of de-ionized water on changes in porosity and permeability of shales mineralogy due to clay-swelling. *Sci. Rep.* **2021**, *11*, 20049. [CrossRef]
2. Wang, L. Clay Stabilization in Sandstone Reservoirs and the Perspectives for Shale Reservoirs. *Adv. Colloid Interface Sci.* **2020**, *276*, 102087. [CrossRef]
3. Zhuang, Y.; Liu, X.; Xiong, H.; Liang, L. Microscopic Mechanism of Clay Minerals on Reservoir Damage during Steam Injection in Unconsolidated Sandstone. *Energy Fuels* **2018**, *32*, 4671–4681. [CrossRef]
4. Sui, W.; Tian, Y.; Yao, C. Investigation of microscopic pore structure variations of shale due to hydration effects through SEM fixed-point observation experiments. *Pet. Explor. Dev.* **2018**, *45*, 894–901. [CrossRef]
5. Lin, H.; Sun, X.; Yuan, Y.; Lai, X.; Qu, H.; Luo, C. Experimental investigation on the dynamic volume changes of varied-size pores during shale hydration. *J. Nat. Gas Sci. Eng.* **2022**, *101*, 104506. [CrossRef]

6. Pang, J.; Wu, T.; Yu, X.; Zhou, C.; Chen, H.; Gao, J. The Effect of Water–Rock Interaction on Shale Reservoir Damage and Pore Expansion. *Processes* **2025**, *13*, 1265. [CrossRef]
7. Zhao, P.; Fan, X.; Zhang, Q.; Yao, B.; Zhang, M.; He, L.; Qiang, Y.; Liu, J. Characteristics of hydration damage and its influence on permeability of lamellar shale oil reservoirs in Ordos Basin. *Geofluids* **2021**, *2021*, 6646311. [CrossRef]
8. Zhou, Y.; You, L.; Kang, Y.; Jia, C.; Xiao, B. Influencing Factors and Application of Spontaneous Imbibition of Fracturing Fluids in Lacustrine and Marine Shale Gas Reservoir. *Energy Fuels* **2022**, *36*, 3606–3618. [CrossRef]
9. Xue, H.; Zhou, S.; Jiang, Y.; Zhang, F.; Dong, Z.; Guo, W. Effects of hydration on the microstructure and physical properties of shale. *Pet. Explor. Dev.* **2018**, *45*, 1146–1153. [CrossRef]
10. Dong, W.; Tian, J.; Chen, Q.; Wang, P.; Han, T.; Zhai, Y.; Jiang, D. Relationship between shale hydration and shale collapse. *ACS Omega* **2022**, *7*, 42524–42536. [CrossRef]
11. Zeng, B.; Dong, E.; Yao, Z.; Song, Y.; Xiong, Z.; Huang, Y.; Gou, X.; Hu, X. Strategies for Optimizing Shut-In Time: New Insights from Shale Long-Term Hydration Experiments. *Processes* **2024**, *12*, 1096. [CrossRef]
12. Zeng, F.; Zhang, Q.; Guo, J.; Zeng, B.; Zhang, Y.; He, S. Mechanisms of shale hydration and water block removal. *Pet. Explor. Dev.* **2021**, *48*, 646–653. [CrossRef]
13. Zhang, G.; Wang, H.; Li, F.; Wang, D.; Li, N.; He, S. Effects of Hydration during Drilling on Fracability of Shale Oil Formations: A Case Study of Da'anzhai Section Reservoir in Sichuan Basin, China. *Processes* **2022**, *10*, 2313. [CrossRef]
14. Li, S.; Zhou, P.; Lan, B. Study of wellbore instability in shale formation considering the effect of hydration on strength weakening. *Front. Earth Sci.* **2024**, *12*, 1403902. [CrossRef]
15. Ma, T.; Chen, P. Study of meso-damage characteristies of shale hydration based on CT scanning technology. *Pet. Explor. Dev.* **2014**, *41*, 227–233. [CrossRef]
16. Wang, X.; Cui, B.; Feng, Z.; Shao, H.; Huo, Q.; Zhang, B.; Gao, B.; Zeng, H. In-situ hydrocarbon formation and accumulation mechanisms of micro- and nano-scale pore-fracture in Gulong shale, Songliao Basin, NE China. *Pet. Explor. Dev.* **2023**, *50*, 1269–1281. [CrossRef]
17. He, W.; Zhu, R.; Cui, B.; Zhang, S.; Meng, Q.; Bai, B.; Feng, Z.; Lei, Z.; Wu, S.; He, K.; et al. The Geoscience Frontier of Gulong Shale Oil: Revealing the Role of Continental Shale from Oil Generation to Production. *Engineering* **2023**, *28*, 79–92. [CrossRef]
18. Meng, S.; Tao, J.; Li, T.; Li, D.; Wang, S.; Yang, L.; Liu, X.; Liang, L.; Liu, H. Mechanical characteristics and reservoir stimulation mechanisms of the Gulong shale oil reservoirs, the northern Songliao Basin. *Pet. Sci.* **2024**, *21*, 2023–2036. [CrossRef]
19. Li, S.; Liang, K.; Wang, C.; Wang, Y.; Jiao, Y.; Zhu, X.; Wang, C. Study of wellbore instability and collapse mechanism for a layered continental shale oil horizontal well. *Energies* **2022**, *15*, 4538. [CrossRef]
20. Liu, Y.; Li, W.; Yuan, J.; Li, P.; Ge, X.; Ge, X.; Liu, P.; Wu, H.; Yu, X.; Huang, B. Lithofacies Characteristics of the Lower Cretaceous Qing 1 Member in the Heiyupao Depression, Northern Binbei Area of the Songliao Basin. *Minerals* **2025**, *15*, 125. [CrossRef]
21. Kang, Y.; Zhu, R.; Liu, K.; Zhang, J.; Liu, C. Pore Fractal and Structure Analysis of Pore-Filling Chlorite in Continental Shales: A Case Study from the Qingshankou Formation in the Gulong Sag. *Fractal Fract.* **2025**, *9*, 266. [CrossRef]
22. Wang, X.; Zhu, X.; Zhang, X. Genetic Mechanisms and Multiparameter Logging Identification of Low-Resistivity Oil Pay: A Case Study of the Triassic Chang 6 Member, Zhidan Area, Ordos Basin, China. *ACS Omega* **2023**, *8*, 23425–23445. [CrossRef] [PubMed]
23. Huo, J.; Zhang, S.; Liu, J.; Wei, C.; Zhang, R.; Zhang, X.; Li, S. Synthesis, Investigation of Temperature and Salt Resistant Polyacrylamide Microspheres Used for Deep Sealing and Profile Control and Function Strengthening Mechanism. *J. Appl. Polym. Sci.* **2024**, *141*, e55620. [CrossRef]
24. Yang, T.; Huang, B.; Zhan, C.; Jiang, C.; Zhang, L.; Zhao, X.; Zhao, M. Mechanical Properties and Mechanisms of Soda Residue and Fly Ash Stabilized Soil. *Sci. Rep.* **2025**, *15*, 1103. [CrossRef]
25. Li, B.; Guo, Y.; Hu, X.; Wang, T.; Wang, R.; Chen, X.; Fan, W.; Deng, Z. Pore Structure and Heterogeneity Characteristics of Deep Coal Reservoirs: A Case Study of the Daning–Jixian Block on the Southeastern Margin of the Ordos Basin. *Minerals* **2025**, *15*, 116. [CrossRef]
26. Wang, T.; Zhou, G.; Fan, L.; Zhang, D.; Shao, M.; Ding, R.; Li, Y.; Hu, H.; Deng, Z. Full-Scale Pore and Microfracture Characterization of Deep Coal Reservoirs: A Case Study of the Benxi Formation Coal in the Daning–Jixian Block, China. *Int. J. Energy Res.* **2024**, *2024*, 5772264. [CrossRef]
27. Shen, R.; Zhang, X.; Yang, H.; Ren, H.; Guo, H.; Zhou, H. Micropore structure and fractal characteristics of shale in Midwest China. *J. Pet. Explor. Prod. Technol.* **2024**, *14*, 867–881. [CrossRef]
28. Jiang, W.; Zhang, Y.; Ma, T.; Chen, S.; Hu, Y.; Wei, Q.; Zhuang, D. Pore Structure and Its Fractal Dimension: A Case Study of the Marine Shales of the Niutitang Formation in Northwest Hunan, South China. *Fractal Fract.* **2025**, *9*, 49. [CrossRef]
29. Dehghanpour, H.; Lan, Q.; Saeed, Y.; Fei, H.; Qi, Z. Spontaneous imbibition of brine and oil in gas shales: Effect of water adsorption and resulting microfractures. *Energy Fuels* **2013**, *27*, 3039–3049. [CrossRef]
30. Zhu, B.; Meng, J.; Song, C.; Pan, R.; Zhu, Z.; Jin, J. Complexity and Heterogeneity Evaluation of Pore Structures in the Deep Marine Shale Reservoirs of the Longmaxi Formation, China. *J. Mar. Sci. Eng.* **2023**, *11*, 1613. [CrossRef]

31. Yang, H.; Shi, X.; Yin, C.; Liang, X.; Zhao, J.; Li, J.; Zhu, J.; Geng, Z.; Wu, Z.; Li, R. Brazilian tensile failure characteristics of marine shale under the hydration effect of different fluids. *Nat. Gas Ind.* **2020**, *40*, 72–80. [CrossRef]
32. Shi, B., Xia, B., Lin, Y., Xu, J. CT imaging and mechanism analysis of crack development by hydration in hard-brittle shale formations. *Acta Pet. Sin.* **2012**, *33*, 137–142. [CrossRef]
33. Chen, M.; Ning, X.; Kang, Y.; Wu, J.; Li, B.; Shi, Y.; Lai, Z.; Bai, J.; Yan, M. Effect of Fracturing Fluid Properties on the Flowback Efficiency of Marine and Continental Transitional Shale Gas Reservoirs in Ordos Basin. *Processes* **2025**, *13*, 1398. [CrossRef]
34. Wang, B.; Liu, B.; Yang, J.; Bai, L.; Li, S. Compatibility characteristics of fracturing fluid and shale oil reservoir: A case study of the first member of Qingshankou Formation, northern Songliao Basin, Northeast China. *J. Pet. Sci. Eng.* **2022**, *211*, 110161. [CrossRef]

minerals

MDPI

Article

Palygorskite as an Extender Agent in Light Cement Pastes for Oil Wells: Performance Analysis

Rafael A. Ventura [1], José V. A. Carvalho [2], Raphael R. da Silva [2], Francisco G. H. S. Pinto [1], Júlio C. O. Freitas [1] and Sibele B. C. Pergher [1,*]

1 Postgraduate Program in Chemistry, Universidade Federal do Rio Grande do Norte, Av. Senador Salgado Filho, 3000, Natal 59072-970, RN, Brazil; rafael.ventura.071@ufrn.edu.br (R.A.V.); gustavo.hayala.099@ufrn.edu.br (F.G.H.S.P.); julio.freitas@ufrn.br (J.C.O.F.)

2 Postgraduate Program in Chemical Engineering, Universidade Federal do Rio Grande do Norte, Av. Senador Salgado Filho, 3000, Natal 59072-970, RN, Brazil; valderisso.carvalho.116@ufrn.edu.br (J.V.A.C.); raphael.ribeiro.089@ufrn.edu.br (R.R.d.S.)

* Correspondence: sibele.pergher@ufrn.br

Abstract: Cementing operations are among the most critical steps in oil-well construction. When performed improperly, the integrity and useful life of the well can be significantly compromised. Light cement pastes are used to cement formations with a low fracture gradient to ensure zonal isolation and maintain the integrity of the casing. Extenders are additives used to reduce the density of cement pastes, ensuring that the paste has desirable properties before and after setting. This work aimed to evaluate the application of palygorskite clay as an additive in lightweight cement pastes for oil wells, highlighting how its fibrous morphology influences the microstructure and enhances the macroscopic properties of the hardened cement matrix. For this, the clay sample was initially characterized regarding its physicochemical properties using X-ray diffraction (XRD), X-ray fluorescence (XRF), thermogravimetry (TG), textural analysis (BET/N_2), and scanning electron microscopy (SEM). Lightweight pastes (1.56 g/cm^3) were then formulated, varying the clay concentration by 1%, 3%, and 6% of the total mass. Cement pastes using bentonite were also formulated for comparison. Technological tests of atmospheric consistency, rheological behavior, free water, and stability were applied. It can be noted that the pastes formulated with palygorskite had lower viscosity, reflected in the reduced plastic viscosity and yield stress values, indicating easier flow behavior when compared with bentonite-based pastes. The pastes formulated with 6% palygorskite and 3% bentonite showed satisfactory stability and drawdown results. Therefore, applying palygorskite satisfies the minimum requirements for acting as an extending agent for lightweight cement pastes and is an option for application in oil-well cementing operations.

Keywords: palygorskite; bentonite; extenders; cement; oil well

1. Introduction

Palygorskite, also known as attapulgite, is a fibrous clay mineral composed of a double layer of silicon tetrahedrons and a central layer of magnesium, aluminum, or iron octahedrons. This substance is found almost exclusively in soils of the world's arid and semiarid regions, with Brazil's primary deposits located in Piauí.

Palygorskite was not initially recognized as a distinct clay mineral when it was discovered due to its chemical composition and similar properties to montmorillonite [1]. However, it stands out as a representative of the group of fibrous clay minerals. Its fibrous morphology is characterized by long, thin fibers with diameters ranging from 20 to

30 nanometers and lengths that can reach several micrometers, whereas montmorillonite, which also has a double layer of silicon tetrahedra and a central layer of aluminum octahedra, exhibits a more lamellar structure [2].

Palygorskite belongs to the palygorskite and sepiolite groups, which are three-layer phyllosilicates with a T:O ratio of 2:1. Furthermore, these phyllosilicates have a fibrous morphology. The physicochemical properties of palygorskite are remarkable, making its applications diverse [3]. With a high specific surface area and a porous structure that varies from micropores to mesopores, palygorskite is an ideal material for applications that require high adsorption capacity such as catalysis, effluent treatment, and as an additive for the petroleum industry [4].

Oil-well cementing is a fundamental process in the construction of oil wells, aiming to ensure the structural stability of the well and the isolation of production zones. The procedure involves pumping a Portland cement slurry between the casing and the rock formation. Proper cementing is essential to ensure the integrity of the well throughout its life, preventing incidents such as blowouts and casing collapses. An effective paste must be adapted to each well scenario; for this purpose, chemical additives are used to achieve certain desired properties [5].

The admixtures used In cement slurries are crucial In modifying their rheological and performance properties. Among the various types of admixtures, extenders are of particular interest as they allow for cement savings and a reduction in the paste density, facilitating their application in easily fractured formations [6]. Extenders, such as clays, increase the volume of the paste without significantly compromising its strength after curing. They are instrumental in situations where it is necessary to minimize fluid invasion into the formation. Palygorskite is already known as a viscosifying additive in the oil industry for its application in drilling fluids [7,8]. This article aimed to investigate palygorskite clay as a potential extender additive for cement pastes used in oil wells. Bentonite clay, widely recognized and commonly used for this purpose, was used as a comparative reference [9].

Although palygorskite has been extensively studied for applications in drilling fluids, its use as an extender in oil-well cementing remains underexplored. Most extenders, such as bentonite, rely on lamellar structures with high swelling capacity but also impose drawbacks like excessive viscosity and reduction in mechanical strength. In contrast, palygorskite presents a fibrous morphology that promotes the formation of a physical network within the cement matrix. This network contributes not only to viscosity modulation, but also enhances the microstructural integrity of the hardened paste by improving the particle packing, reducing pore size, and mitigating sedimentation. This study sought to bridge this knowledge gap by correlating the microstructural characteristics induced by palygorskite with the macroscopic performance of cement pastes, offering an alternative extender with improved operational properties for oil-well cementing.

2. Materials and Methods

2.1. Materials

The materials used to develop this research were Portland cement Class G (density: 3.13 g/cm^3) from Mizu Cimentos LTDA (Mossoro, Brasil), bentonite (density: 2.34 g/cm^3) from Bentonisa S/A (João Pessoa, Brasil) and palygorskite (density: 2.42 g/cm^3) from the state of Piaui, Brazil.

It is important to note that the performance of clay minerals such as palygorskite and bentonite can vary considerably depending on their geological origin and mineralogical composition. The findings of this study were based on a specific sample, and while they provide meaningful insights into the comparative behavior of these materials, the results should not be generalized to all types of palygorskite or bentonite without further validation.

2.2. Clay Characterization

The palygorskite and bentonite, previously crushed in an agate mortar, were subjected to the following physicochemical characterization techniques: X-ray diffraction (XRD), X-ray fluorescence spectroscopy (XRF), scanning electron microscopy (SEM), thermogravimetric analysis (TGA/DTG), and the adsorption and desorption of nitrogen at 77 K. The XRD analysis was performed using Bruker (Billerica, MA, USA) D2Phaser equipment equipped with a Lynxeye detector and copper radiation (CuKα, λ = 1.54 Å) with a Ni filter, current of 10 mA, voltage of 30 kV, 2-theta range between 2° and 40°, divergent slit of 0.6 mm, central slit of 1 mm, step equal to 0.01°, and acquisition time of 0.2 s. The methods of powder (natural sample) and oriented blade (hydrated, glycolated, and calcined samples) were used.

The XRF analysis was performed using energy-dispersive X-ray fluorescence (EDX) equipment on a Bruker (Billerica, MA, USA) S2 PUMA—SERIES II device, using a Pd tube with a maximum power of 50 W, maximum voltage of 50 kV, maximum current of 1 mA and a HighSense Silicon Drift Detector.

The thermal stability of the clay was studied by thermogravimetry using a TG 209 F3 Tarsus—Netzsch thermal analyzer (Weimar, Germany). Samples of about 8 mg were placed in platinum crucibles and heated from 26 to 900 °C at a heating rate of 10 °C/min in an inert nitrogen atmosphere with a flow rate of 50 mL/min. Scanning electron microscopy (SEM) analysis was performed on a TESCAN (Brno, Czech Republic) VEGA 4 scanning electron microscope using a secondary electron (SE) detector with a beam energy of 10 KeV. The samples were coated with gold film and deposited on carbon strips for analysis in a high vacuum.

The N_2 adsorption and desorption analysis at 77 K was carried out using the NOVA 800 BET equipment from Anton Paar (Graz, Austria). The samples were previously degassed at 150 °C for 10 h. With the data collected, graphs of the adsorption and desorption isotherms were plotted. From the data obtained, the specific surface area was determined using the BET method, which followed the criteria suggested by IUPAC [10], and the total pore volume using the Gurvith method.

2.3. Technological Tests for Evaluating the Extender Agent for Cement Slurry

Initially, the formulations were calculated according to the API RP 10B-2 standards [11], which define a volume of 600 cm^3 of cement slurry for carrying out technological tests. In addition, the density was set at 13.0 lb/gal (1.56 g/cm^3). Seven formulations were defined, with the masses described in Table 1: a standard formulation containing only water and cement and six slurries with additives of palygorskite and bentonite at 1%, 3%, and 6% by weight of cement. Due to its high consistency, the paste with 6% bentonite (B6) was not adequately mixed according to the requirements of API RP 10B-2 [11], making it not feasible to carry out the technological tests.

It is important to emphasize that the inability to prepare the slurry with 6% bentonite (B6) according to API RP 10B-2 [11] was not solely a laboratory limitation, but rather an indication of the excessive viscosifying effect of bentonite at higher concentrations. During the preparation procedure, the slurry reached a consistency level that exceeded the capacity of the mixing equipment and deviated from the operational standards established for slurry homogenization.

Table 1. Cement slurry designs used in the tests.

Slurry	Water (g)	Cement Class G (g)	Bentonite (g)	Palygorskite (g)
P0	440.76	493.87	-	-
B1	439.98	489.74	4.90	-
B3	438.46	481.71	14.45	-
B6	436.28	470.13	28.21	-
P1	439.79	489.93	-	4.90
P3	438.83	486.07	-	14.47
P6	436.99	469.46	-	28.17

The slurries were prepared according to the procedure recommended in API RP 10B-2 [11]. All components were weighed on an electronic scale, ensuring an accuracy of 0.1% of the masses, and the mixture was mixed in a Chandler (Tulsa, OK, USA) brand paddle mixer, Model 3500, with speed and mixing time control. The bentonite and palygorskite were previously hydrated in the mixing water for 30 min at a rotation of 1000 rpm. The prepared slurries were subjected to atmospheric consistency, free water, stability, rheology, and compressive strength tests.

The atmospheric conditions for the tests were defined using the Schedule 2010 software, version 1.6. Considering a well with a depth of 1000 m and a geothermal gradient of 1.7 °F/100 ft, the following temperatures were established: bottomhole circulation temperature (BHCT) of 38 °C and bottomhole static temperature (BHST) of 57 °C.

Before measuring the atmospheric consistency and performing the free water test, the cement slurries were homogenized using a Chandler atmospheric consistometer, model 1200, operating at a rotation of 150 ± 15 rpm for 30 min at room temperature. After the homogenization period, the consistency of the slurries was assessed with a consistency measuring dial. Then, approximately 623 g of the paste was transferred to a 500 mL Erlenmeyer flask, which remained at rest for 2 h. The supernatant water was removed with a syringe and weighed. Finally, the percentage of free water present in the slurry was calculated.

In the stability test, the slurries were previously homogenized at 38 °C (simulating BHCT). After 30 min of homogenization, they were poured into decanting tubes and placed in a thermostatic bath, where they remained for 24 h. After being removed from the bath, the cylinders were opened at the top, and the existing fluids were eliminated. The remaining space up to the top of the cylinder was filled using a syringe, obtaining a volume in cubic centimeters (cm^3). The "top reduction" was obtained by converting the volume to length, expressed in mm. Subsequently, the hardened slurry was removed from the cylinders and sectioned into four equal parts. Based on Archimedes' principle, the difference in density between the top and bottom sections was verified.

The rheological test was performed using a Chandler viscometer, model 3500. The slurries, previously homogenized at a temperature of 38 °C, were subjected to variable rotation rates (ascending and descending) from 3 to 300 rpm, recording the respective torque values. Finally, the rheological parameters of the initial gel, final gel, yield point, and plastic viscosity were obtained using the Bingham mathematical model.

The compressive strength test was performed in triplicate. The slurries were placed in cubic metal molds with 50.00 mm edges and cured for 7 and 28 days in a thermostatic bath, under agitation, at a temperature of 57 °C. The rupture of the test specimens was performed using a Shimadzu (Kyoto, Japan) AG-I 300 kN universal mechanical testing machine controlled by Trapezium X software. The compression tests were performed at room temperature with a loading speed of 72.2 kN/min.

3. Results and Discussion

3.1. Characterization of Palygorskite Clay

3.1.1. X-Ray Diffraction (XRD)

X-ray diffraction analysis of the palygorskite sample was performed in four ways. Initially, an XRD analysis of the untreated material was performed. Next, a clay hydration procedure was performed, following the oriented blade methodology. The third analysis consisted of a glycolization process using ethylene glycol to monitor any significant changes in the diffractogram. Finally, the sample was calcined at a temperature of 600 °C, and then XRD analysis was performed. Figure 1 shows the diffractograms obtained for the four analyses performed with the palygorskite sample.

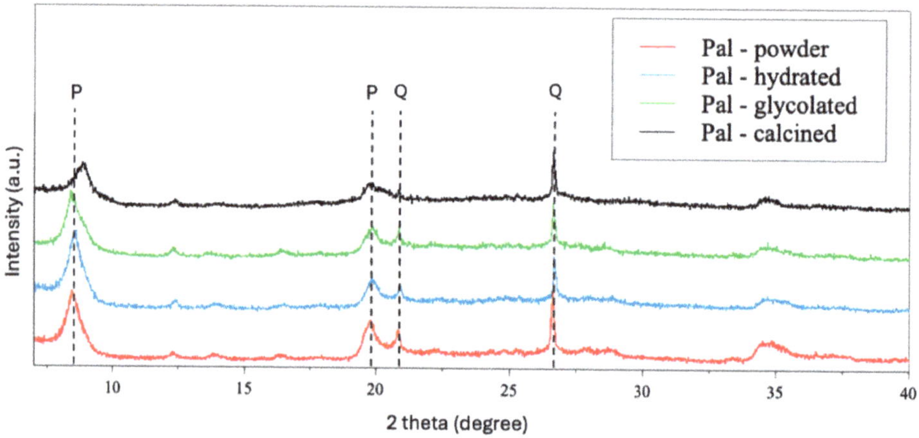

Figure 1. Diffractogram of palygorskite clay under different conditions.

As can be seen, the diffractograms of the four analyses were similar, presenting only a slight change in the intensity of the peaks. In agreement with the literature [12,13], one can observe four prominent peaks in the diffractogram above. The first refers to the palygorskite itself, in the order of $2\theta = 8.4°$, followed by a peak in the order of $2\theta = 20°$, also from palygorskite. The two peaks at $2\theta = 21°$ and $2\theta = 27°$ refer to quartz, which was also present in this clay sample, as is well-referenced in other works [14].

The X-ray diffraction (XRD) patterns of sodium bentonite subjected to distinct physicochemical treatments—powder, hydrated, glycolated, and calcined—revealed a structural behavior typical of smectite-group minerals, particularly montmorillonite, as shown in Figure 2.

In its powder form, the (001) basal reflection was observed at approximately $2\theta \approx 6.8°$, corresponding to a basal spacing of about 13 Å, which is characteristic of natural sodium montmorillonite with ambient moisture. Upon hydration, this reflection shifts slightly toward lower angles, indicating an expansion of the interlayer spacing due to water intercalation. When glycolated, the (001) peak moved significantly to lower 2θ values ($\approx$5.6–6.0°), reaching a basal spacing of around 17 Å—a diagnostic indicator of expandable smectites such as montmorillonite. Conversely, the calcined sample exhibited a substantial reduction in the (001) peak intensity and the collapse of the layered structure, consistent with dehydroxylation and partial amorphization resulting from high-temperature treatment. The appearance of minor peaks suggests the persistence of thermally stable impurities such as quartz.

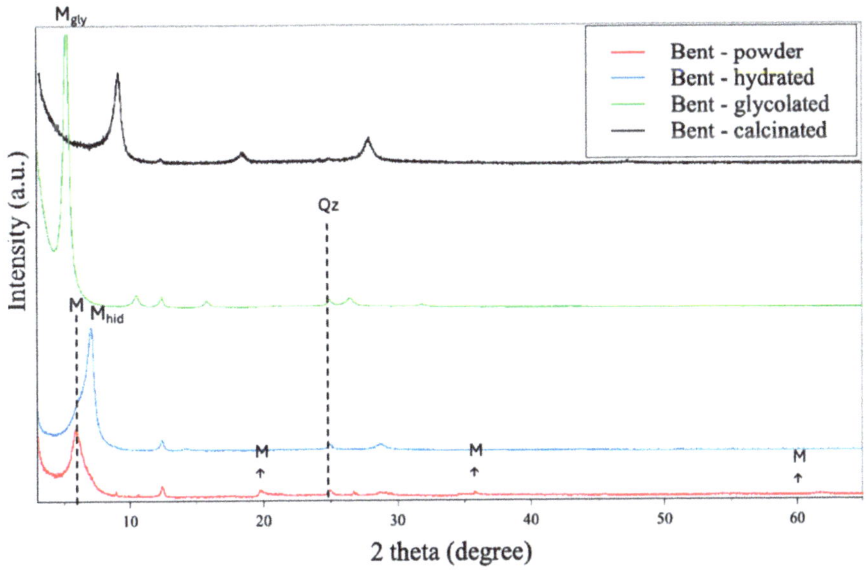

Figure 2. Diffractogram of bentonite clay under different conditions.

This sequential shift of the (001) peak under glycolation and collapse under calcination confirmed the dominant presence of sodium montmorillonite and its characteristic responsiveness to hydration, swelling, and thermal degradation, in agreement with established clay mineralogical behavior [15].

3.1.2. X-Ray Fluorescence (XRF)

XRF analysis was performed using energy-dispersive X-ray fluorescence (EDX) equipment. Table 2 shows the results obtained from the equipment.

Table 2. Chemical characterization of the palygorskite and bentonite samples.

Oxide	% Weight	
	Palygorskite	**Bentonite**
MgO	5.35	3.81
Al_2O_3	18.36	23.85
SiO_2	59.78	56.62
K_2O	3.69	0.55
CaO	0.58	1.26
TIO_2	1.08	0.63
Fe_2O_3	11.16	13.28

The results are in line with the chemical analysis of palygorskite and bentonite, showing high values in the quantity of silicon, aluminum, and iron. Another metal that stood out in these samples was the presence of magnesium. These data are well-referenced in the literature [16–18].

3.1.3. Scanning Electron Microscopy (SEM)

Below are two SEM images, Figure 3, of the palygorskite sample captured at 10,000× and 20,000× magnification and 10 KeV electron beam energy. The detector used was the secondary electron (SE).

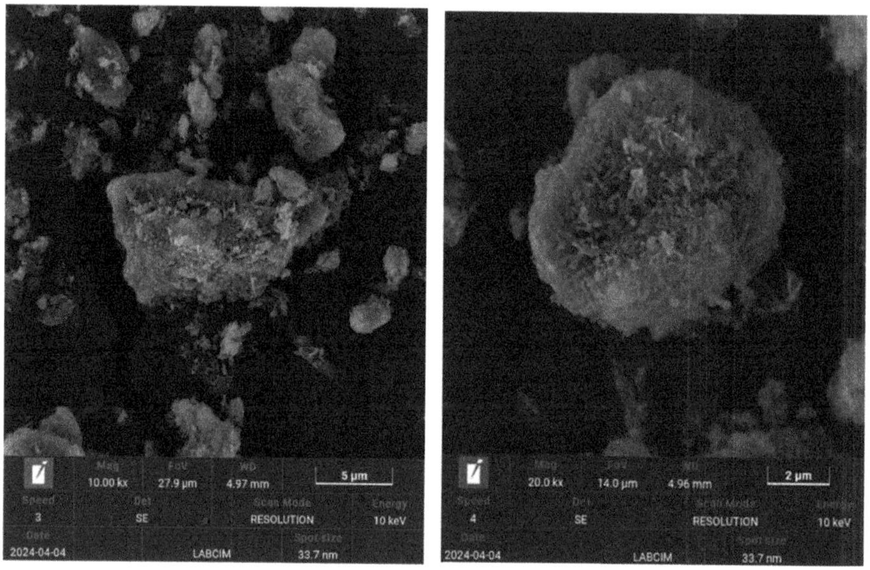

(**a**) (**b**)

Figure 3. SEM images of the palygorskite sample: (**a**) image of a palygorskite particle at 10,000× magnification; (**b**) image of another palygorskite particle at 20,000× magnification.

The micrograph above is characteristic of palygorskite; it is possible to see a fibrous structure and a coiling of these fibers. These images align with what can be observed in the literature [16]. Bentonite was analyzed using the same equipment and with the same image acquisition parameters, as can be seen in Figure 4.

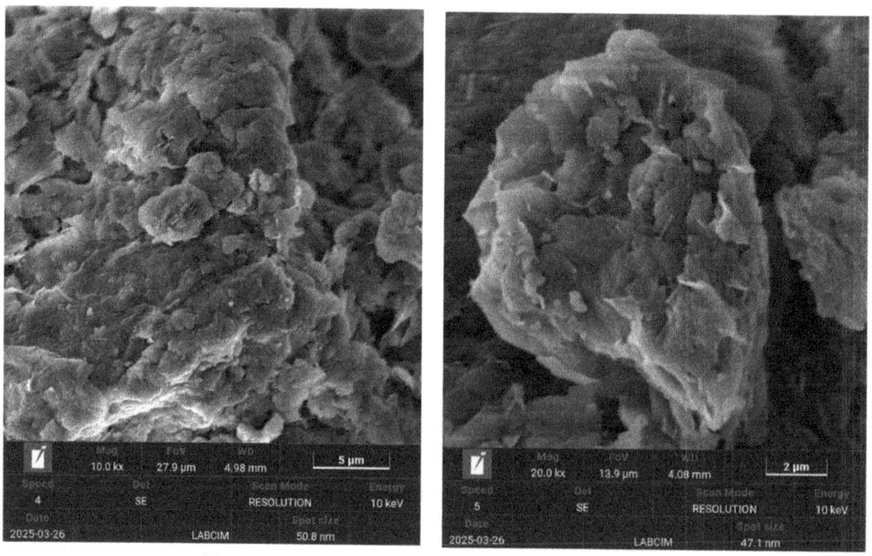

(**a**) (**b**)

Figure 4. SEM images of the bentonite sample: (**a**) image of a bentonite particle at 10,000× magnification; (**b**) image of another bentonite particle at 20,000× magnification.

The scanning electron microscopy (SEM) image of the bentonite sample revealed a characteristic lamellar aggregated structure, commonly associated with smectite group minerals, particularly montmorillonite. The particles displayed a plate-like or flaky morphology, forming stacked and irregularly arranged aggregates. This structural organization is consistent with the typical expandable layered nature of montmorillonite [19].

3.1.4. Thermogravimetry (TG)

Thermogravimetric analysis was performed using NETZSCH TG 209F3 equipment. A heating ramp up to 900 °C was programmed with a heating rate of 10 °C/min. Figure 5 shows the TG curves of the palygorskite and bentonite samples analyzed.

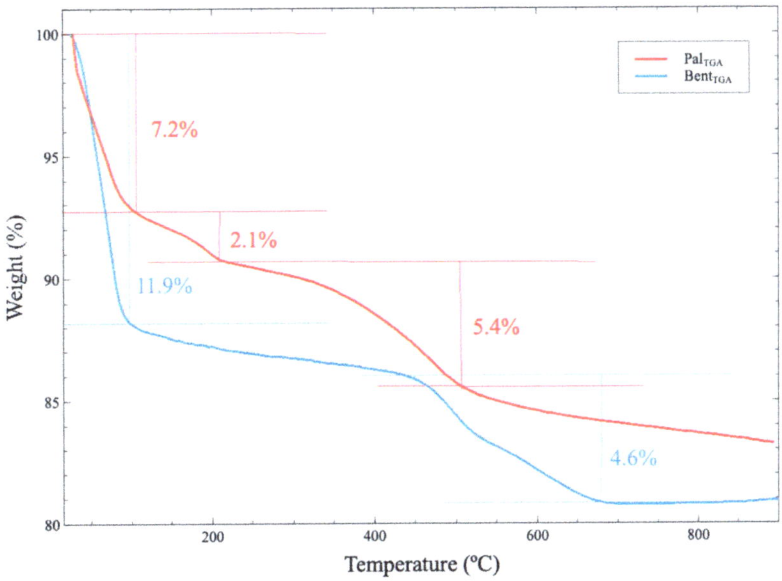

Figure 5. TG curves of the palygorskite and bentonite samples.

In the red curve (palygorskite), the three significant losses were related to water. The 7% mass loss, which occurred up to 105 °C, was attributed to water physisorbed on the palygorskite surface and micropore water [20]. The second peak of 2% mass loss, between 120 °C and 240 °C, was attributed to the loss of the remaining micropore water and the loss of the first coordination waters. Finally, the last peak of 5% mass loss, which occurred up to 569 °C, referred to the loss of residual coordination waters and the loss of structural waters [21].

The thermogravimetric behavior of bentonite, represented by the blue curve, revealed a significant initial mass loss of 11.9%, primarily attributed to the release of physically adsorbed and interlayer water. This phenomenon is characteristic of the smectite group, particularly montmorillonite, which constitutes the dominant mineral phase in bentonite. This dehydration process typically occurs at temperatures up to approximately 150 °C and reflects the high water retention capacity of bentonite due to its expansive lamellar structure [22]. A second mass loss of 4.6% was observed between 450 °C and 670 °C, corresponding to the dehydroxylation of structural hydroxyl groups within the clay layers. This thermal behavior is consistent with previous studies that emphasized the thermal response of montmorillonite-based clays under progressive heating [23].

When comparing bentonite and palygorskite, it was evident that bentonite underwent a more pronounced initial mass loss, indicative of its higher adsorption capacity, which is a direct consequence of its expandable smectitic structure. In contrast, palygorskite exhibited a more gradual and distributed thermal response due to its fibrous morphology and internal channels that retain water in various forms (adsorbed, channel, and structural).

3.1.5. Nitrogen Adsorption at 77 K

The palygorskite and sodium bentonite samples were analyzed for the adsorption and desorption of N_2 at 77 K. Important information on the textural properties of both materials was obtained, and Figure 6 shows the graph of the results.

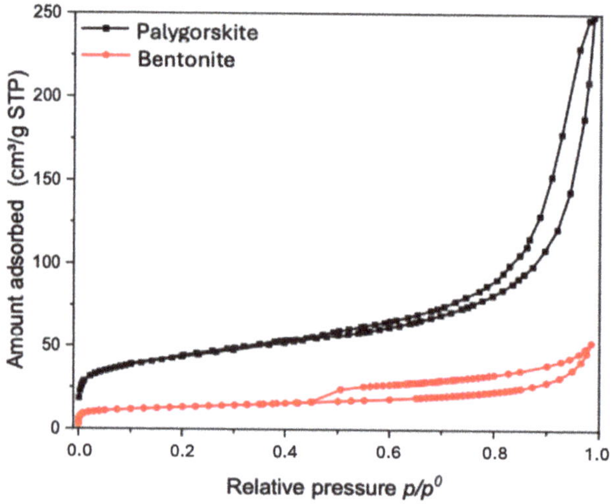

Figure 6. Adsorption and desorption of N_2 at 77 K.

The plotted graph showed a curve similar to a type II isotherm with slight hysteresis, type H3, for the palygorschite sample, while for sodium bentonite, the isotherm was type II with hysteresis type H4. It is worth remembering that the difference in the types of hysteresis of the materials is due to the geometry of the pores of the materials. The presence of hysteresis is indicative of the presence of mesopores [24]. The area data were calculated from the BET equation following the criteria suggested by IUPAC.

After making the equation of the line and calculating the angular and linear coefficients, it was possible to calculate the constant C and the capacity of the monolayer (N) to then apply the final formula for the specific area and obtain the value of 142.59 m^2/g [11]. The literature offers various specific area values for palygorskite, but all of the values are as high as the one calculated experimentally [25]. The properties of sodium bentonite and palygorskite are shown in the Table 3.

Table 3. The N_2 adsorption and desorption characteristics of palygorskite and bentonite.

Material	SBET (m^2/g)	Constante C	VTP (cm^3/g)	Vmeso(cm^3/g)
Palygorskite	142.59	218.28	32.75	29.25
Bentonite	47.05	344.12	0.08	0.073

3.2. Technological Tests for Evaluating Extenders for Cement Slurries

3.2.1. Atmospheric Consistency

After mixing the slurries added with palygorskite or bentonite as an extending agent and the standard slurry, their consistency was verified at room temperature and pressure. Figure 7 shows the results after 30 min of testing.

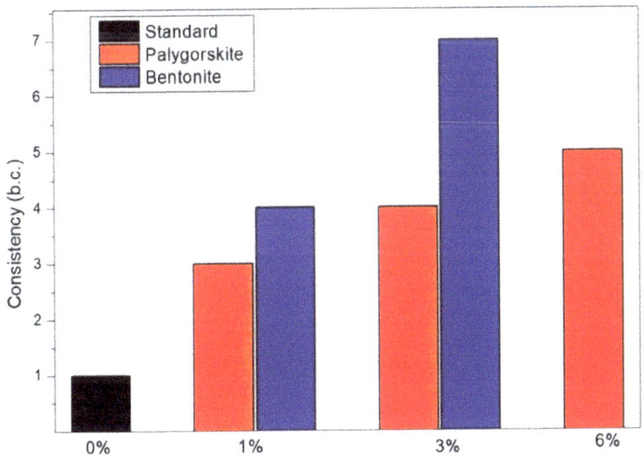

Figure 7. Consistency behavior of slurries at room temperature and pressure.

It can be observed that the standard slurry (only cement and water) presented a low consistency, which characterized it as a light paste. An increase in consistency was observed with the rising concentration of palygorskite. The cement paste containing 3% palygorskite, compared with the standard paste (0%), exhibited a consistency increase from 1 b.c. to 4 b.c. A similar behavior was observed for bentonite, where the consistency reached 7 b.c. in the paste with the 3% addition.

Bentonite clay has a 2:1 structure, that is, an octahedral layer between two tetrahedral layers (as can be seen in Figure 8), and in the interlayers, there are exchangeable cations, which give them a high swelling capacity, and therefore a high swelling aptitude. This behavior offered a higher viscosity when compared with the other pastes [26]. The inability to prepare the B6 slurry (6% bentonite) due to excessive consistency demonstrates a key operational limitation of bentonite at high concentrations. Its strong swelling behavior causes an exponential viscosity increase, making the slurry unworkable. In contrast, palygorskite maintained a workable consistency even at 6%.

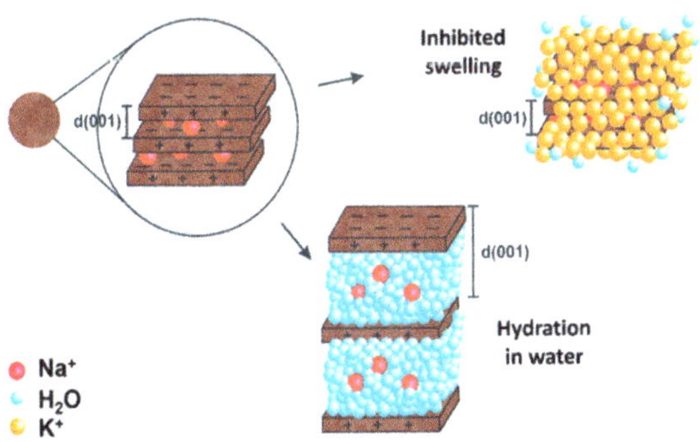

Figure 8. Hydration and swelling inhibition scheme of sodium bentonite clay.

3.2.2. Free Water

The free water content refers to the amount of supernatant water, that is, water that does not participate in the cement hydration reactions. The amount of free water is a determining factor for the integrity and durability of the cementing job. A low amount of free water is crucial to ensure the strength and stability of the cement paste, avoiding problems such as cracks and loss of adhesion. According to the API standard [11], free water is an essential parameter for slurries intended for cementing oil wells. It cannot exceed 5.9% of the volume of the paste prepared with Class G Portland cement. Table 4 presents the free water results for the standard paste and the slurries with palygorskite and bentonite.

Table 4. Free water results of the formulated pastes.

Standard	P1	P3	P6	B1	B3
18.62%	16.69%	12.61%	4.25%	5.08%	2.10%

The standard slurry presented a high percentage of free water, much higher than the 5.9% recommended by the API [11]. On the other hand, it was found that the addition of both clays promoted a reduction in this percentage. Concentrations of 1% and 3% of bentonite were sufficient to reduce the free water in the paste to levels below the maximum recommended by the standard. Concentrations of 1% and 3% of palygorskite resulted in higher percentages than 5.9% of free water. However, with the addition of 6% palygorskite, it was possible to reduce the free water content to 4.25%.

These results indicate that the addition of palygorskite, especially at appropriate concentrations, can be an effective strategy to control the free water content, similar to bentonite. This contributes to optimizing the cement slurry properties, ensuring a more efficient and safer cementing job.

3.2.3. Rheology

After being conditioned in an atmospheric consistometer for 30 min at a temperature of 38 °C, the rheological behavior of the slurries was verified using a Chandler viscometer, model 3500. The instrument is equipped with coaxial cylinder geometry, featuring a bob with a diameter of 34.49 mm and a slit size of 1170 mm. The ascending and descending readings were taken from 3 to 300 rpm, respecting 10 s after each speed. Figure 9 shows the flow curves of the formulated pastes.

All of the analyzed pastes presented non-Newtonian behavior, with a decrease in viscosity and an increase in the shear rate. The Birgham model presented a good fit for the curves, with coefficients of determination greater than 0.99. The standard paste, P0, and the paste, P1, presented the lowest shear stress values with the increase in the shear rate, with a gradual increase with the increase in the addition of palygorskite clay to the mixture. The addition of bentonite clay provided a more significant increase in viscosity to the pastes, consistent with what was observed in the atmospheric consistency behavior [27].

Figure 10 presents the results of the rheological parameters of the formulated pastes' yield point (a), plastic viscosity (b), and gel strength (c), where Gi represents the initial gel and Gf the final gel.

The yield strength of a cement paste is a critical property in the area of oil- and gas-well cementing. It refers to the stress at which the slurry begins to deform and flow, indicating its ability to start flowing under stress, which is essential for pumping operations. A higher yield strength helps to suspend solid particles uniformly in the paste, preventing sedimentation and ensuring uniform distribution. An adequate yield strength helps to achieve good zonal isolation, ensuring that the paste fills the annular space and maintains its integrity until it hardens [28]. By analyzing Figure 10a, it was possible to notice that bentonite clay performed better when compared with the other formulations. This advantage is directly related to its hydration and swelling capacity due to the exchangeable cations in its interlayers.

The same observation can be said for both the parameters of plastic viscosity (Figure 10b) and gel strength (Figure 10c). However, although presenting even lower values than bentonite, palygorskite clay with an addition higher than 3% presented a better rheological performance when compared with the standard paste.

The lower plastic viscosity and yield point observed in slurries containing palygorskite, compared with those with bentonite, are directly related to the distinct morphology of the additive. While bentonite exhibits a lamellar structure with high swelling capacity due to interlayer water adsorption, palygorskite is composed of rigid, rod-like fibrous particles that do not swell but form a physical entangled network within the slurry. This fibrous arrangement Increases the slurry's structural integrity without significantly increasing the viscosity. From an operational perspective, this characteristic translates into improved pumpability and lower frictional pressure losses during displacement. The reduced viscosity facilitates the more efficient removal of drilling fluids from the annulus, potentially enhancing the zonal isolation quality.

However, while lower viscosity improves the flow properties, it may pose challenges regarding the suspension of heavier particles during static periods. Nevertheless, the rheological parameters observed—particularly the yield stress and gel strength for formulations with 3% and 6% palygorskite—indicate that the slurry maintains an adequate suspension capability, balancing flowability with stability.

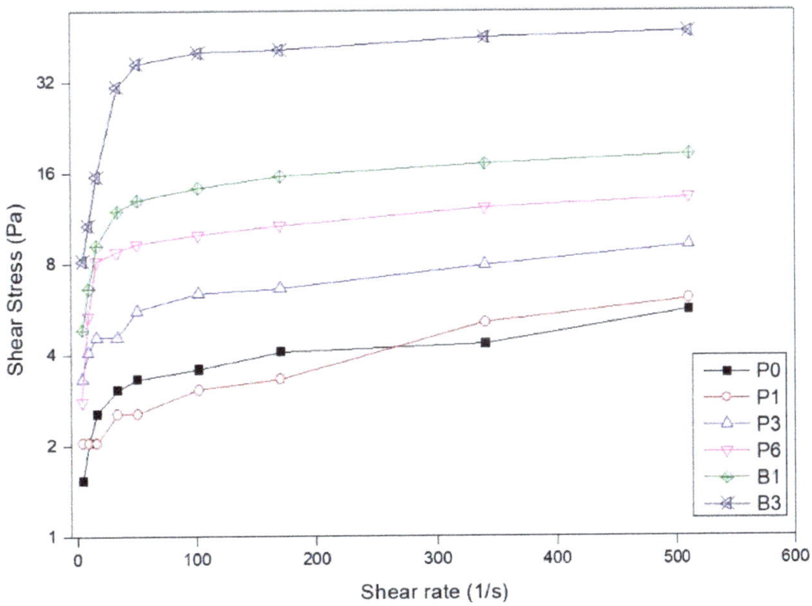

Figure 9. Flow curves of the formulated pastes.

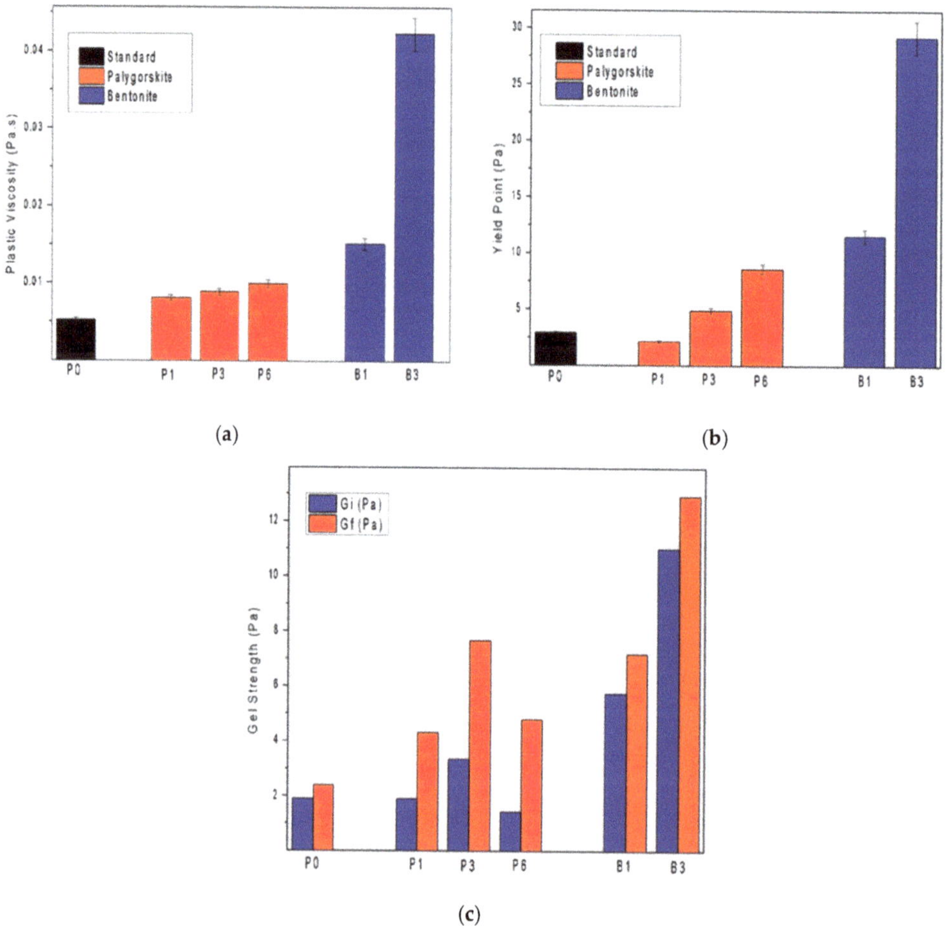

Figure 10. Rheological parameters of the formulated pastes: (**a**) plastic viscosity, Pa; (**b**) yield limit, Pa.s; (c) gel strength, Pa.

3.2.4. Stability Test

Figures 11 and 12 show the density variation data and the drawdown of the formulated slurries after one day of curing. According to API RP10 [11], if the bottom and top density differences are more significant than 0.5 lb/gal (0.06 g/cm^3), or the top's drawdown is superior to 5 mm, the paste is considered unstable and must be reformulated.

Only slurries P6 and B3 presented a satisfactory difference in density between the top and bottom within the limits established by the standard. This is because the concentration of dispersed solids was small, which resulted in less interaction between particles, causing the solids to settle. Silva et al. [29] studied cement pastes with bentonite added as an extender, varying between 12 lb/gal (1.44 g/cm^3) and 13.5 lb/gal (1.62 g/cm^3), and observed that the lighter pastes presented less stability when compared with those with higher density, having a more significant drawdown. Bentonite was more effective in pastes with higher density, 13.5 lb/gal (1.62 g/cm^3), and clay concentrations greater than 3%.

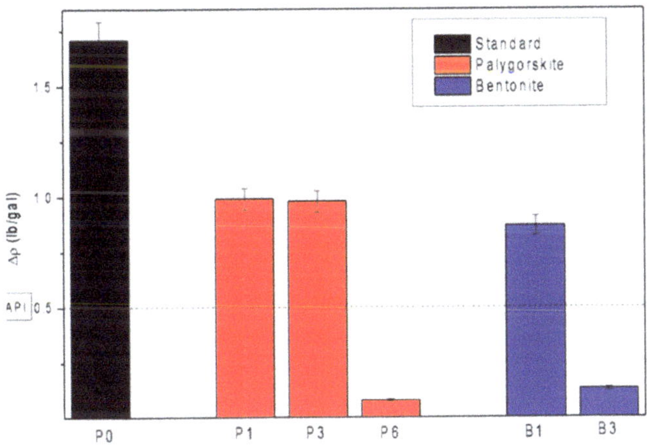

Figure 11. Stability of the formulated pastes.

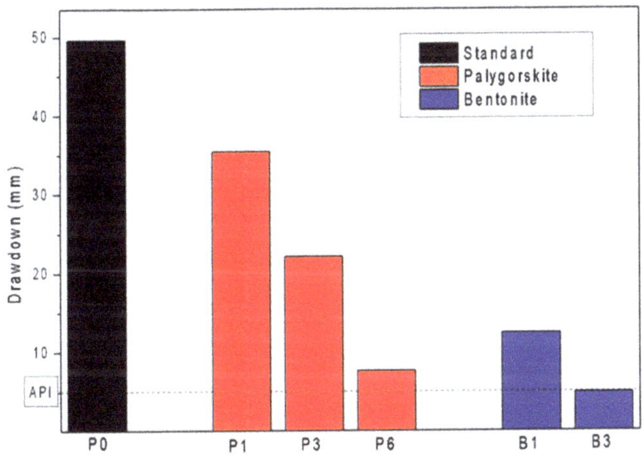

Figure 12. Drawdown of the formulated pastes after 24 h of curing.

The stability results observed for slurries containing 6% palygorskite and 3% bentonite were closely related to their ability to control free water. In the case of bentonite, stability was mainly achieved through water absorption into its interlayer spaces, characteristic of smectite clays with high swelling capacity. In contrast, palygorskite stabilized the slurry via a distinct mechanism: its fibrous morphology promoted the formation of an interwoven physical network within the cementitious matrix. This fiber-based network mechanically trapped water within the capillary spaces between fibers and enhanced particle retention, effectively reducing sedimentation and limiting free water migration toward the surface.

This structural reinforcement explains why the slurry containing 6% palygorskite exhibited a significant reduction in free water (4.25%) and maintained a uniform density profile within the stability thresholds defined by the API standard. The physical entrapment of water, coupled with the scaffolding effect provided by the fibrous network, contributed to enhanced homogeneity and structural integrity of the slurry during both the fluid phase and subsequent setting.

3.2.5. Compressive Strength

The evaluation of the mechanical strength of the cement slurries through a compression test is essential to ensure the full functioning of the paste in an oil well. The paste must have sufficient resistance to ensure the physical stability of the well and prevent cracks that could compromise its safety. The slurries were cured for 7 and 28 days at 57 °C for the uniaxial compression strength test. The pastes selected for this test were the standard paste, B1 (bentonite 1%), B3 (bentonite 3%), P1 (palygorskite 1%), and P6 (palygorskite 6%). Figure 13 shows the maximum stress results supported by each sample in the compression test.

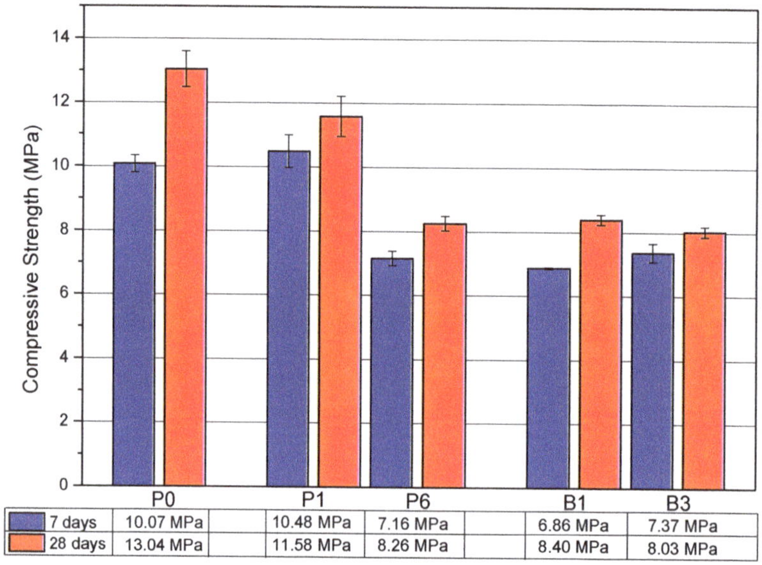

Figure 13. Compressive strength of the formulated pastes after curing for 7 and 28 days at a temperature of 57 °C.

One of the main factors in the search for a replacement for bentonite is its negative influence on the mechanical strength of the slurry. When comparing the standard slurry with those containing 1% and 3% bentonite, a decrease of approximately 35.58% and 38.42% in strength at 28 days was observed. This reduction may be critical for the inadequacy of these formulations in cementing projects that require high mechanical strength values [30]. Among the slurries studied, P1, containing 1% by mass of palygorskite, presented the best performance, having an increase of 4.07% in strength for the 7-day period. By analyzing the 28-day period, there was a decrease of only 11.19% in mechanical strength, a value three times lower compared with the same concentration of bentonite at 28 days. Adding an extender to the cementitious matrix resulted in a decrease in cementitious particles, and consequently a drop in mechanical strength [31]. This phenomenon was observed when adding more palygorskite to the paste and increasing the bentonite concentration.

The compressive strength behavior of the slurries containing palygorskite indicates that its influence extends beyond the simple dilution of the cementitious phase. The increased strength observed in the P1 slurry (1% palygorskite) can be attributed to a microstructural reinforcement mechanism, where the fibrous particles act as physical bridges within the cement matrix. This effect likely enhances particle interlocking, improves matrix cohesion, and may refine the pore structure, reducing the initiation and propagation of microcracks.

However, at higher concentrations (6%), the excessive amount of fibers appeared to disrupt the continuity of the cementitious matrix. This oversaturation can lead to weak interfacial zones, limit the full development of hydration products, and introduce additional porosity due to the presence of excessive non-reactive material.

3.3. *Microstructural Analysis of Hardened Cement Paste Containing Palygorskite and Bentonite*

Scanning electron microscopy (SEM) analysis was performed on the hardened cement pastes containing palygorskite and bentonite to evaluate the influence of their distinct morphologies on the cement matrix.

The SEM images of the paste containing palygorskite (Figure 14) clearly revealed the presence of a fibrous network embedded within the cementitious matrix. The palygorskite fibers were distributed throughout the structure, forming an interlaced network that appeared to contribute to the physical cohesion of the matrix. This fibrous arrangement likely acts as a reinforcing framework, bridging hydration products, filling capillary spaces, and potentially mitigating the development of microcracks. The fibers also seemed to contribute to a denser and more homogeneous microstructure, which correlated with the enhanced stability and improved rheological behavior observed in the macroscopic tests.

(a) (b)

Figure 14. SEM micrographs of the hardened cement paste containing palygorskite at (**a**) 2000× and (**b**) 5000× magnification.

In contrast, the SEM image of the paste containing bentonite (Figure 15) displayed a markedly different microstructure. The morphology was dominated by aggregated clusters of lamellar particles characteristic of smectite-type clays. These plate-like structures were less integrated into the cementitious matrix compared with palygorskite. The bentonite clusters tended to form localized porous zones, where the layered arrangement may have trapped water and promoted microvoid formation. This microstructural feature likely contributed to the higher free water content and lower overall matrix cohesion observed in the bentonite-based slurry formulations.

Figure 15. SEM micrograph of the hardened cement paste containing bentonite at 2000× magnification.

These microstructural observations strongly support the rheological and mechanical performance trends discussed earlier. The fibrous morphology of palygorskite provides not only viscosity control and stability benefits in the fresh slurry, but also contributes to the physical integrity of the hardened cement matrix.

4. Conclusions

The main objective of this study was to evaluate the effectiveness of palygorskite clay as an extending agent in lightweight cement pastes for application in oil-well cementing operations, comparing it with the agent most commonly used by the industry, bentonite. Some conclusions can be drawn based on the results of palygorskite characterization and technological application.

- Palygorskite has proven to be a very versatile clay. As it is a 2:1 phyllosilicate, its specific area is large, providing essential industry applications. The nitrogen adsorption analysis allowed for the calculation of the specific area, which resulted in 142 m^2/g, which is a high specific area value, and is one of its significant advantages.
- As it does not have exchangeable cations in its interlayers, after treatments, no changes were observed in the diffractogram patterns obtained for the palygorskite, with only slight increases in intensity in some peaks.
- The addition of palygorskite in light cement pastes influenced the rheological behavior, presenting lower viscosities when compared with pastes added with bentonite;
- The results obtained from the rheological parameters of the bentonite slurries were superior, given their more significant swelling capacity after hydration.
- The pastes with 6% palygorskite and 3% bentonite were those that presented superior stability, showing no sedimentation and providing better applicability in cementing operations.
- Additionally, the rheological behavior of the palygorskite-based slurries demonstrated a relevant operational advantage. The lower viscosity resulting from its fibrous morphology improved the pumpability and displacement efficiency, reducing the risk of high frictional pressures during cementing operations. At the same time, the

physical entanglement of fibers ensured a sufficient yield strength and gel strength to maintain particle suspension, achieving a balance between flowability and stability.

- The paste with 1% palygorskite presented a superior result to the pastes with the addition of bentonite when their compressive strengths were evaluated.
- The microstructural evaluation confirmed that palygorskite, due to its fibrous morphology, promoted the formation of a more interconnected and homogeneous cement matrix. This characteristic contributed significantly to enhancing the mechanical integrity of the hardened cement paste. These findings position palygorskite as a technically viable alternative to traditional extenders, with additional microstructural benefits that extend beyond a simple density reduction.

Author Contributions: Conceptualization, R.A.V.; Methodology, R.A.V., J.V.A.C. and R.R.d.S.; Validation, J.C.O.F.; Formal analysis, R.A.V. and F.G.H.S.P.; Investigation, R.A.V., J.V.A.C. and R.R.d.S.; Resources, J.C.O.F. and S.B.C.P.; Data curation, F.G.H.S.P.; Writing—original draft preparation, R.A.V., J.V.A.C. and R.R.d.S.; Writing—review and editing, S.B.C.P.; Supervision, J.C.O.F.; Project administration, J.C.O.F.; Funding acquisition, S.B.C.P. All authors have read and agreed to the published version of the manuscript.

Funding: This research received no external funding.

Data Availability Statement: Data are contained within the article.

Conflicts of Interest: The authors declare no conflicts of interest.

References

1. Murray, H.H. Traditional and New Applications for Kaolin, Smectite, and Palygorskite: A General Overview. *Appl. Clay Sci.* **2000**, *17*, 207–221. [CrossRef]
2. Sun, X.; Chen, Y.; Liang, L.; Xie, G.; Peng, Y. Research on Hydrocyclone Separation of Palygorskite Clay. *Minerals* **2023**, *13*, 1264. [CrossRef]
3. Pan, M.; Su, J.; Tang, L.; Hu, Z.; Huang, X. Insights into Phosphate Adsorption Performance onto Magnetic Thermal Modified Palygorskite Nanocomposites. *Minerals* **2023**, *13*, 293. [CrossRef]
4. Baltar, C.A.M.; da Luz, A.B.; Baltar, L.M.; de Oliveira, C.H.; Bezerra, F.J. Influence of Morphology and Surface Charge on the Suitability of Palygorskite as Drilling Fluid. *Appl. Clay Sci.* **2009**, *42*, 597–600. [CrossRef]
5. Murthy, R.V.V.R.; Mohammad, F.; Chavali, M. Development of Innovative Lightweight Slurry in Oil Well-Cementing Operations. *Upstream Oil Gas Technol.* **2021**, *7*, 100045. [CrossRef]
6. Adjei, S.; Elkatatny, S. Overview of the Lightweight Oil-Well Cement Mechanical Properties for Shallow Wells. *J. Pet. Sci. Eng.* **2021**, *198*, 108201. [CrossRef]
7. Santanna, V.C.; Araújo, G.C.N.; da Silva MT, A.; de Castro Dantas, T.N.; Pimentel, P.M. Water-Based Drilling Fluid with Palygorskite: Cutting Carrying and Contaminants. *Clay Miner.* **2023**, *58*, 95–101. [CrossRef]
8. Zhuang, G.; Zhang, J.; Chen, J.; Liu, Q.; Fan, W.; Li, Q. Application of Nanofibrous Clay Minerals in Water-Based Drilling Fluids: Principles, Methods, and Challenges. *Minerals* **2024**, *14*, 842. [CrossRef]
9. Magzoub, M.I.; Shamlooh, M.; Salehi, S.; Hussein, I.; Nasser, M.S. Gelation Kinetics of PAM/PEI Based Drilling Mud for Lost Circulation Applications. *J. Pet. Sci. Eng.* **2021**, *200*, 108383. [CrossRef]
10. Thommes, M.; Kaneko, K.; Neimark, A.V.; Olivier, J.P.; Rodriguez-Reinoso, F.; Rouquerol, J.; Sing, K.S.W. Physisorption of Gases, with Special Reference to the Evaluation of Surface Area and Pore Size Distribution (IUPAC Technical Report). *Pure Appl. Chem.* **2015**, *87*, 1051–1069. [CrossRef]
11. API. API SPEC 10A—Cements and Materials for Well Cementing | GlobalSpec. Available online: https://standards.globalspec.com/std/14208303/api-spec-10a (accessed on 11 January 2025).
12. Liu, Y.; Wang, W.; Wang, A. Effect of Dry Grinding on the Microstructure of Palygorskite and Adsorption Efficiency for Methylene Blue. *Powder Technol.* **2012**, *225*, 124–129. [CrossRef]
13. Zhang, T.; Huang, X.; Qiao, J.; Liu, Y.; Zhang, J.; Wang, Y. Recent Developments in Synthesis of Attapulgite Composite Materials for Refractory Organic Wastewater Treatment: A Review. *RSC Adv.* **2024**, *14*, 16300–16317. [CrossRef] [PubMed]
14. Frost, R.L.; Xi, Y.; He, H. Synthesis, Characterization of Palygorskite Supported Zero-Valent Iron and Its Application for Methylene Blue Adsorption. *J. Colloid Interface Sci.* **2010**, *341*, 153–161. [CrossRef]

15. Moore, D.M.; Reynolds, R.C. *X-Ray Diffraction and the Identification and Analysis of Clay Minerals*, 2nd ed.; Oxford University Press: New York, NY, USA, 1997.
16. Xavier, K.C.M.; Filho, E.C.S.; Santos, M.S.F.; Santos, M.R.M.C.; Luz, A.B. Mineralogical, Morphological, and Surface Characterization of Attapulgite from Guadalupe-PI. *Holos* **2012**, *5*, 60–70. [CrossRef]
17. Liu, H.; Xie, B.; Qin, Y. Effect of Bentonite on the Pelleting Properties of Iron Concentrate. *J. Chem.* **2017**, *2017*, 7639326. [CrossRef]
18. Tsakiri, D.; Douni, I.; Taxiarchou, M. Structural and Surface Modification of Oxalic-Acid-Activated Bentonites in Various Acid Concentrations for Bleaching Earth Synthesis—A Comparative Study. *Minerals* **2022**, *12*, 764. [CrossRef]
19. Christidis, G.E. Industrial Clays. In *Handbook of Clay Science*, 2nd ed.; Bergaya, F., Lagaly, G., Eds.; Elsevier: Amsterdam, The Netherlands, 2013; Volume 5, pp. 213–236.
20. Önal, M.; Sarıkaya, Y. Some Physicochemical Properties of a Clay Containing Smectite and Palygorskite. *Appl. Clay Sci.* **2009**, *44*, 161–165. [CrossRef]
21. Wang, S.; Liu, X.; Zhang, C.; Hu, W.; Liu, Y.; Fu, X.; Jun, Y.; Wei, S. Adsorption and Selective Mechanism of Pb2+ and Cd2+ on the Surface of Calcined Modified Attapulgite. *Sep. Purif. Technol.* **2025**, *353*, 128377. [CrossRef]
22. Sakizci, M.; Erdoğan Alver, B.; Yörükoğullari, E. Thermal Behavior and Immersion Heats of Selected Clays from Turkey. *J. Therm. Anal. Calorim.* **2009**, *98*, 429–436. [CrossRef]
23. Madejová, J.; Komadel, P. Baseline Studies of the Clay Minerals Society Source Clays: Infrared Methods. *Clays Clay Miner.* **2001**, *49*, 410–432. [CrossRef]
24. Zhang, J.; Wang, Q.; Chen, H.; Wang, A. XRF and Nitrogen Adsorption Studies of Acid-Activated Palygorskite. *Clay Miner.* **2010**, *45*, 145–156. [CrossRef]
25. Chang, Y.; Liu, H.; Zha, F.; Chen, H.; Ren, X.; Lei, Z. Adsorption of Pb(II) by N-Methylimidazole Modified Palygorskite. *Chem. Eng. J.* **2011**, *167*, 183–189. [CrossRef]
26. Adjei, S.; Elkatatny, S.; Sarmah, P.; Abdelfattah, A.M. Evaluation of Calcined Saudi Calcium Bentonite as Cement Replacement in Low-Density Oil-Well Cement System. *J. Pet. Sci. Eng.* **2021**, *205*, 108901. [CrossRef]
27. Nascimento, R.C.A.M.; Amorim, L.V.; Santana, L.N.L. Development of aqueous fluids with bentonite clay for drilling of onshore oil wells. *Cerâmica* **2010**, *56*, 179–187. [CrossRef]
28. Li, P.; Hu, M.; Liu, M.; Zhang, H.; Zhao, J.; Xia, X.; Guo, J. Regulation of Oil Well Cement Paste Properties by Nanoclay: Rheology, Suspension Stability and Hydration Process. *Constr. Build. Mater.* **2023**, *377*, 131049. [CrossRef]
29. Silva, F.P.F.; Marques, J.F.S.; Vieira, M.M.; Melo, M.A.F.; Freitas, J.C.O.; Santos, P.H.S.; Dantas, T. OTIMIZAÇÃO DE FORMULAÇÃO DE PASTAS DE CIMENTO CONTENDO BENTONITA PARA APLICAÇÃO EM POÇOS PETROLÍFEROS COM BAIXO GRADIENTE DE FRATURA. In Proceedings of the 8° Congresso Brasileiro de Pesquisa e Desenvolvimento em Petróleo e Gás. Curitiba, Brazil, 20–22 October 2015.
30. Fode, T.A.; Jande, Y.A.C.; Kivevele, T. Effects of Raw and Different Calcined Bentonite on Durability and Mechanical Properties of Cement Composite Material. *Case Stud. Constr. Mater.* **2024**, *20*, e03012. [CrossRef]
31. Li, Z.; Tang, L.; Zheng, Y.; Tian, D.; Su, M.; Zhang, F.; Ma, S.; Hu, S. Characterizing the Mechanisms of Lead Immobilization via Bioapatite and Various Clay Minerals. *ACS Earth Space Chem.* **2017**, *1*, 152–157. [CrossRef]

minerals

MDPI

Article

Analyzing the Mechanism of Drilling Losses in the Zhengning Oilfield in the Ordos Basin

Tuan Gu [1], Shugang Yang [1], Yunfeng Xiao [1], Linpeng Zhang [1], Fangquan Peng [2], Xu Su [2,*], Tao Fan [1], Haiyang Wang [2] and Desheng Zhou [2]

[1] Exploration and Development Research Institute of Liaohe Oilfield Company, PetroChina, Panjin 124010, China; gutuan@petrochina.com.cn (T.G.); 13679161464@163.com (S.Y.); xiaoyunf@petrochina.com.cn (Y.X.); zlp0303@petrochina.com.cn (L.Z.); fantao414@163.com (T.F.)

[2] School of Petroleum Engineering, Xi'an Shiyou University, Xi'an 710065, China; 23212010116@stumail.xsyu.edu.cn (F.P.); wanghaiyang@xsyu.edu.cn (H.W.); dzhou@xsyu.edu.cn (D.Z.)

* Correspondence: 24111010032@stumail.xsyu.edu.cn

Abstract: Frequent wellbore loss incidents in the Mesozoic reservoirs of the Zhengning oilfield in the Ordos Basin, China, have severely impacted the development of tight oil and gas reservoirs in the basin. This study analyzed the mineral composition, microstructure, natural fracture distribution, and hydration–dispersion characteristics of rocks in the loss-prone intervals of the Mesozoic reservoirs using a laboratory experimental system. The effects of natural fractures and drilling fluid immersion on the mechanical properties and failure behavior of the rocks were investigated, and the wellbore loss mechanisms in the Mesozoic reservoirs of the Zhengning oilfield were comprehensively analyzed. Experimental results show that the reservoir rocks in the loss-prone intervals are widely distributed, with natural fractures having a width of 0.2–0.3 mm, and the clay mineral content is generally above 40%. When the relative content of illite/smectite interstratification exceeds 80%, the rock exhibits strong hydration–dispersion behavior, with a thermal recovery rate of less than 85%. Drilling fluid immersion causes the rock mass to become unstable. The presence of natural fractures and the hydration–dispersion effect significantly weaken the strength of the surrounding rock mass, leading to enhanced rock plasticity. During drilling, the expansion and interconnection of natural fractures, combined with the hydration–dispersion effect, are the main causes of wellbore loss incidents. Our study clarifies the mechanisms underlying wellbore loss incidents in the Zhengning oilfield and provides reliable experimental evidence for preventing such incidents in this area.

Keywords: Ordos Basin; drilling; rock mineral composition; rock mechanics

1. Introduction

In the Ordos Basin, China's Zhengning oilfield is a key area rich in tight oil and gas resources, so it is crucial to China National Petroleum Corporation (CNPC)'s development strategy [1–3]. The region holds substantial potential, with estimates indicating large reserves vital for national energy security. However, drilling efficiency is hindered by recurrent lost circulation incidents, which increase operational costs and disrupt drilling timelines, challenging timely resource development [4,5]. Lost circulation impacts wellbore stability, formation integrity, and drilling safety [6–8]. Understanding its mechanisms is essential to identify contributing factors and develop effective mitigation strategies [9,10]. This knowledge is critical for optimizing drilling performance, ensuring safety, and sup-

 https://doi.org/10.3390/min15010085

porting sustainable operations in the Zhengning oilfield. Thus, addressing drilling losses is a priority to enhance resource extraction efficiency and reliability in this key region.

Current research on drilling loss mechanisms is analyzed through two primary approaches: numerical simulations and experimental studies. Numerical simulations have significantly advanced, enabling researchers to model the complex behavior of drilling fluids under various geological conditions [11–13]. These simulations primarily focus on stress variations around the wellbore, highlighting how increased pore pressure and stress concentrations can lead to fracture propagation. Fractures may form due to the interaction between the drilling fluid and the rock, leading to significant fluid loss, especially in formations with pre-existing weaknesses [14–17]. Experimental research complements these numerical findings by examining the physical properties of rocks and drilling fluids under controlled conditions [18–20]. Studies have shown that the hydration and dispersion of certain rock types can cause wellbore instability, weakening the rock and potentially leading to collapse [21,22]. For example, clays tend to expand when exposed to water-based drilling fluids, increasing permeability and fluid loss. Experimental setups also allow for the investigation of additional factors, such as temperature variations, fluid viscosity, and the effects of different additives in drilling mud. These factors can either mitigate or exacerbate lost circulation [23–25]. Recent advancements have introduced hybrid approaches that combine numerical and experimental methods, offering a more comprehensive understanding of lost circulation phenomena [26,27]. Research has also emphasized the importance of geological factors, such as natural fractures, faults, and the anisotropic nature of rock materials, which complicate the prediction and management of drilling losses [28–30]. Additionally, the influence of drilling parameters, such as weight on bit, rotation speed, and fluid properties, has been extensively studied to develop best practices for minimizing lost circulation incidents. Overall, this multifaceted approach highlights the need for a deep understanding of the mechanisms behind drilling losses, which is crucial for improving drilling efficiency and safety in challenging environments.

Research has also been conducted on efficient drilling for unconventional oil and gas reservoirs in the Ordos Basin, China. Wu et al. [31] proposed a digital DPM method that can monitor the relative variation of rock mechanical strength along the depth direction of a drill bit in real time. Liu et al. [32] studied the optimal trajectory design for horizontal wells in fractured tight sandstone reservoirs of the Ordos Basin. Zhang et al. [33] developed a novel drilling fluid system with enhanced plugging for carbonate reservoirs in the Ordos Basin. Zhang et al. [34] analyzed the mechanism of formation damage caused by wellbore losses and optimized the drilling fluid system based on their findings. Although some research achievements have been made in efficient drilling in the Ordos Basin, the drilling loss mechanisms and influencing factors in the Zhengning oilfield block of the Ordos Basin remain unclear.

To address the frequent lost circulation issues in the tight oil and gas reservoirs of the Zhengning oilfield, this study first analyzed the mineral composition, natural fracture distribution, and micro-pore throat characteristics of its lost circulation intervals using core samples. Based on this analysis, the hydration and dispersion properties of the reservoir rocks in these intervals were investigated. Finally, utilizing standard cores from the same lost circulation intervals, we examined the mechanical properties of the surrounding rock under various conditions. This research elucidates the mechanisms of lost circulation during drilling in the Zhengning oilfield, providing valuable theoretical guidance for the safe and efficient exploitation of tight oil and gas reservoirs in the Ordos Basin of China.

2. Analysis of Rock Characteristics in Loss Zones

Based on the characteristics of the drilling loss incident at the Zhengning oilfield, we conducted a study and analysis of the drilling loss mechanism in the area following the steps and sequence outlined in Table 1.

Table 1. The steps involved in the drilling loss analysis and their sequence.

Steps for Analyzing Drilling Losses	Experiment Type	Experimental Purpose
Rock characteristic analysis	(1) Steady-state displacement experiment (2) XRD diffraction analysis (3) Scanning electron microscopy (SEM) experiment	Analyze the permeability, porosity, mineral composition, and distribution size of natural fractures in the rock.
Hydration–dispersion characteristic analysis	(1) Rock water absorption experiment (2) Thermal recovery experiment	Analyze the water absorption capacity of different types of rocks and study the structural changes in the rock mass after water absorption.
Rock mechanical characteristic analysis	Triaxial compression experiment	Analyze the mechanical properties of rocks and study the impact of natural fractures and drilling fluid immersion on the mechanical performance of the rock.

Understanding the basic characteristics of reservoir rocks is fundamental for analyzing how losses occur during the drilling process. In the Ordos Basin's Zhengning oilfield, significant loss incidents frequently arise during drilling. To investigate the causes and mechanisms of drilling losses, we conducted core sampling (as shown in Figure 1) to analyze the rock characteristics of loss zones at different depths. Standard cylindrical rock samples with a length of 5.0 cm and a diameter of 2.5 cm were prepared using a wire-cutting machine. We tested the water displacement permeability and porosity of the cores from the loss zones using a steady-state displacement experimental setup, with the results summarized in Table 2. The experimental procedure was as follows: (1) Place the standard cylindrical rock sample in the lithology holder. (2) Turn on the displacement pump and inject distilled water into the rock pores at a constant pressure. (3) Once the flow rate at the outlet stabilizes, record the pressure at both ends of the holder and the flow rate at the outlet. (4) Calculate the porosity and water displacement permeability based on Darcy's law and the mass change of the core before and after displacement. As shown in Table 2, the rocks in the loss zones exhibit low porosity and permeability, with an average porosity of 3.89% and an average permeability of 0.044 mD. This indicates that, during drilling, the drilling fluid has difficulty flowing into the rock pores of the loss zones, which affects the pore pressure and effective stress field surrounding the wellbore.

Using XRD technology, we analyzed the main mineral content of rocks in the loss zones, with the results shown in Figure 2. The experimental procedure was as follows: (1) Select representative core samples and grind them into rock powder using a grinding machine. (2) Analyze the mineral composition of the samples using X-ray diffraction (XRD) equipment. The data indicate that the clay mineral content in the reservoir rocks of the loss zones is generally high, with the clay mineral content consistently exceeding 40% at different depths and reaching up to 70%. Quartz is the second most abundant mineral, averaging around 30%. The contents of potassium feldspar and plagioclase are quite low, collectively accounting for less than 10%. When clay mineral content is high, minerals such as montmorillonite and illite, which are prone to hydration and dispersion, may affect the mechanical properties of the rock under the invasion of drilling fluids.

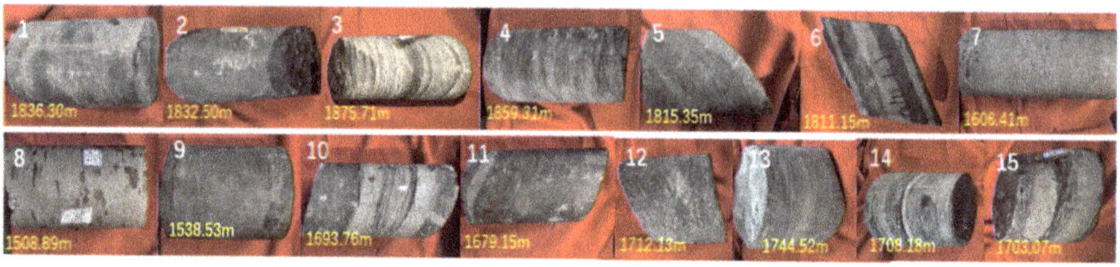

Figure 1. Core sampling during drilling at the Zhengning oilfield.

Table 2. Fundamental physical properties of major loss zone rocks.

Core Number	Porosity/%	Permeability/mD	Depth of Drilling Leakage Zone/m
1	4.76	0.090	1832.5
2	5.10	0.120	1836.3
3	3.83	0.030	1859.31
4	1.98	0.009	1875.71
5	5.35	0.064	1811.15
6	3.69	0.040	1815.35
7	3.22	0.050	1508.89
8	5.76	0.035	1538.53
9	3.35	0.040	1606.41
10	4.15	0.031	1679.15
11	2.36	0.020	1693.76
12	5.18	0.071	1712.13
13	3.31	0.026	1744.52
14	3.96	0.030	1703.07
15	2.34	0.010	1708.18

Based on this, we conducted further analysis of the clay mineral composition in the loss zone rocks, with the results displayed in Figure 3. The data reveal that the content of illite/montmorillonite interlayers in the clay minerals is generally high, averaging over 38%, with some samples approaching 90%. Illite is the second most abundant mineral component in the clay minerals, averaging around 30%. The contents of kaolinite and chlorite are relatively low. The illite/montmorillonite interlayer represents a transitional layer between the layered structures of illite and montmorillonite, exhibiting mixed characteristics in chemical composition and crystal structure. Given that montmorillonite is prone to hydration and dispersion, a high content of illite/montmorillonite interlayers suggests that the clay minerals in the loss zones may possess a certain degree of hydration and dispersion capability.

During drilling, natural fractures in the rock can become the primary channels for drilling fluid loss. Under the pressure of the fluid column in the wellbore, these natural fractures may further propagate, exacerbating the loss of drilling fluid. As shown in Figure 4, during the preparation of standard rock samples, we observed that the rocks in the loss zones contain natural fractures of varying sizes. The natural fractures in the reservoir of the loss zones are predominantly longitudinal. Using a vernier caliper, we measured the widths of the macro natural fractures present in the loss zone reservoir rocks, and the statistical results are presented in Figure 5. The data indicate that the width of the natural fractures has a wide distribution range, with the majority falling within the 0.2 mm to 0.3 mm range, accounting for over 30%. Natural fractures wider than 0.5 mm comprise as much as 14% of the total. During drilling, macro natural fractures with widths greater than 0.1 mm in the reservoir of the loss zones can easily begin to propagate under the static fluid column pressure in the wellbore, leading to drilling fluid loss. Once the fractures begin to propagate, both their width and length can increase rapidly, resulting in larger-scale drilling fluid loss incidents.

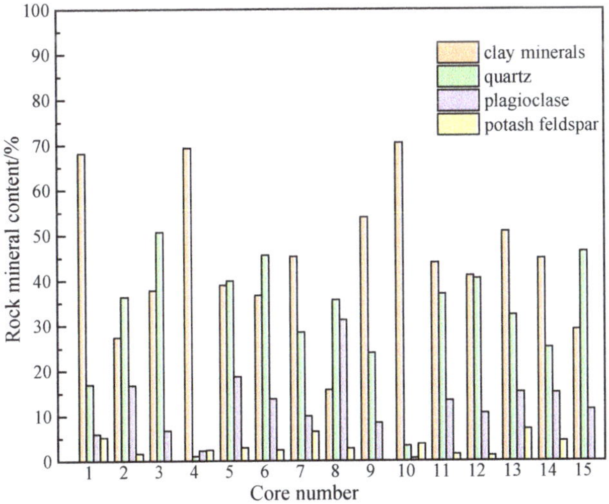

Figure 2. Analysis of main mineral content in reservoir rocks of loss zones.

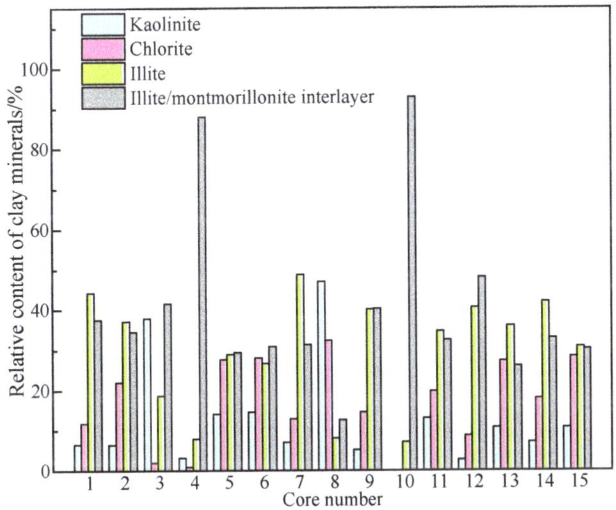

Figure 3. Analysis of relative content of clay minerals in loss zone rocks.

Additionally, we utilized scanning electron microscopy (SEM) to analyze the micropore structure of the rocks in the loss zones, with the results shown in Figure 6. The experimental procedure was as follows: (1) Select drilling core samples from different loss zones. (2) Use a cutting machine to cut the rock into samples of 0.5 cm × 0.5 cm. (3) Clean the surface of the samples and coat them with a thin layer of metal to enhance their conductivity. (4) Use scanning electron microscopy (SEM) to observe the microstructure of the rock surface and the morphology of natural fractures, and then measure the fracture width of the natural fractures. The data indicate that micro fractures in the loss zone rocks are also well developed. The fibrous clay minerals exhibit a flaky structure with a certain degree of orientation, interspersed with quartz crystals. Feldspar particles and fibrous clay mineral particles contain micro fractures of varying widths. We conducted a statistical analysis of the widths of the micro natural fractures in the loss zone rocks, with the results presented in Figure 7. The analysis reveals that the widths of the micro natural fractures primarily range from 0.5 μm to 2 μm, accounting for over 60%, while those exceeding 3.0 μm are relatively rare. During drilling, once the drilling fluid invades the rock pores, it may cause the micro natural fractures to continue to propagate, forming larger fractures. The continued propagation of micro natural fractures during drilling fluid loss not only reduces the strength of the rock surrounding the wellbore but may also connect with larger fractures, resulting in more significant fluid loss. Thus, in the context of the Zhengning oilfield in the Ordos Basin, it is essential to incorporate sealants that can block micro fractures during drilling to prevent the propagation of micro fractures into larger ones. Overall, the analyses in Figures 4–7 suggest that the presence of large macro natural fractures and micro natural fractures in the loss zone rocks is one of the causes of drilling fluid loss incidents. During drilling, large natural fractures directly provide pathways for fluid loss, and under the pressure of the fluid column in the wellbore, these macro natural fractures may continue to propagate and connect with micro natural fractures, resulting in even larger fluid loss.

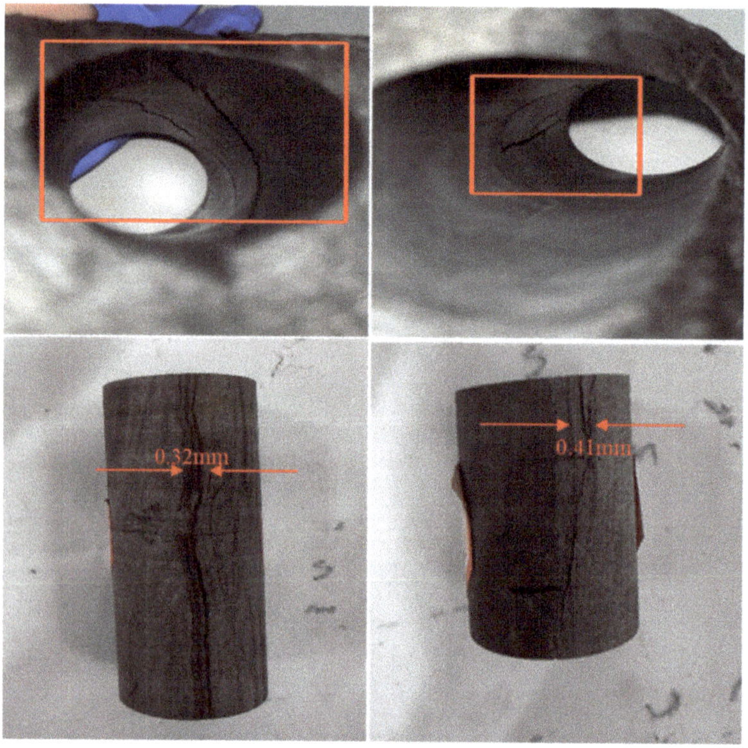

Figure 4. Distribution of macro natural fractures in reservoir rocks of loss zones.

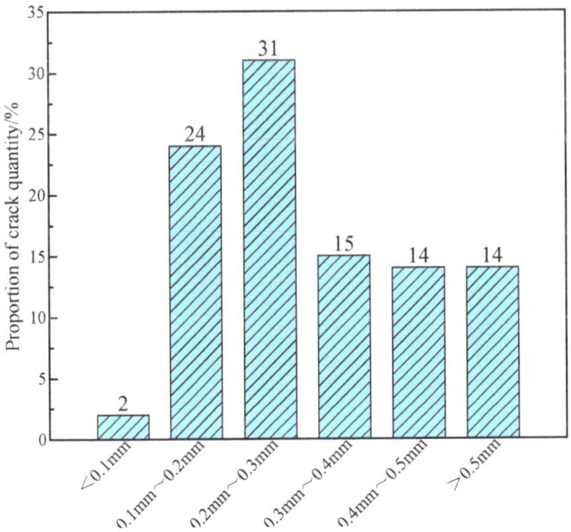

Figure 5. Statistical results of macro natural fracture widths in loss zone rocks.

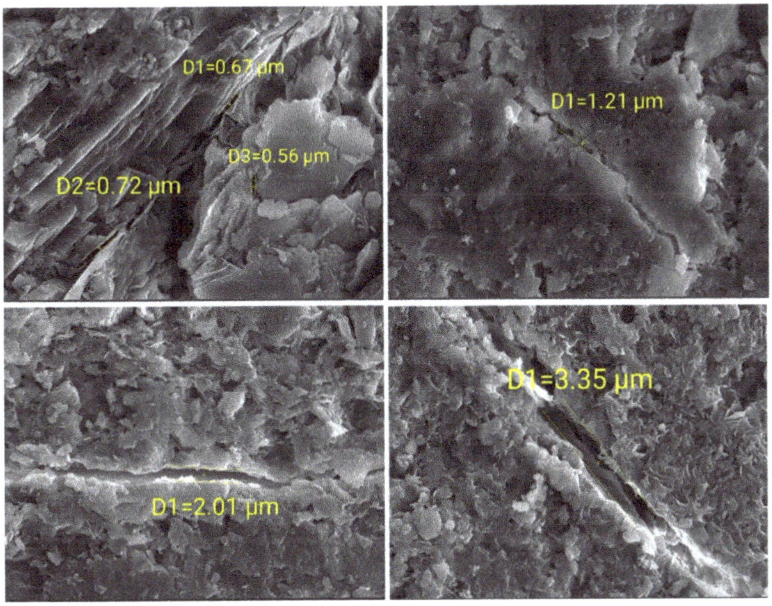

Figure 6. Microstructural features and distribution of micro natural fractures in reservoir rocks of loss zones.

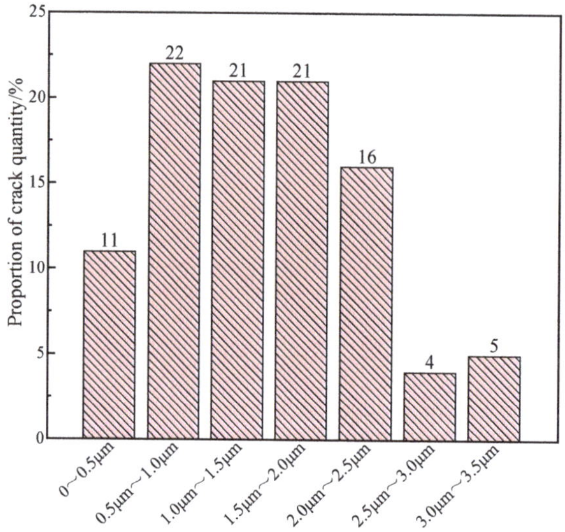

Figure 7. Statistical results of micro natural fracture widths in reservoir rocks of loss zones.

3. Analysis of Hydration and Dispersion Characteristics of Rocks in Loss Zones

In the previous section, we analyzed the mineral composition of rocks in the loss zones, confirming that illite/montmorillonite interlayers are a primary component of clay minerals, which may induce hydration and dispersion in the rocks, affecting their mechanical properties. For this section, we first conducted water absorption experiments to analyze the short-term water absorption performance of four rock samples (core numbers 3, 4, 10, and 12) with a high clay mineral content. The experimental procedure was as follows: (1) Dry the prepared rock samples, record their mass, and measure their volume. (2) Suspend the cores in distilled water and monitor the mass changes during the absorption process. (3) After 48 h of soaking, remove the cores and measure their volume. Figure 8 presents the mass change curve over time for the rock samples immersed in distilled water. Figure 8 shows the curve of mass change over time for the rock samples immersed in distilled water. The results indicate that when the relative content of illite/montmorillonite interlayers in the clay minerals is below 50% (cores 3 and 12), the water absorption curve is relatively flat and the overall absorption is low. Conversely, when the relative content exceeds 80% (cores 4 and 10), the rocks exhibit rapid water absorption, with a significantly larger total absorption amount.

Figure 9 illustrates the hydration and dispersion behavior of the rock samples numbered 3, 4, 10, and 12 after long-term immersion in distilled water. The experimental procedure was as follows: (1) Select drilling core samples with identical quality and intact rock structure. (2) Soak the selected drilling core samples in a container filled with distilled water, ensuring that the entire rock sample is submerged. (3) Observe the changes in the rock structure as the core samples are soaked in distilled water for different durations. As shown in Figure 9, rocks with a low content of illite/montmorillonite interlayers (cores 3 and 12) exhibited no significant structural changes even after being immersed for 20 days, indicating weak hydration and dispersion capabilities. In contrast, the rocks with a high content of illite/montmorillonite interlayers (cores 4 and 10) began to show structural changes after 10 days of soaking, with noticeable chipping and flaking occurring. As the soaking time increased, the hydration and dispersion effects intensified, leading to significant alterations in the rock structure and resulting in instability. We further compared the main and side views of rock sample 10, which had the highest content of illite/montmorillonite interlayers, before and after soaking, as shown in Figure 10. Before immersion, core 10 displayed distinct horizontal bedding fractures with narrow widths. However, after immersion in distilled water, the rock structure underwent significant changes, with a marked increase in the width of the horizontal bedding fractures. Additionally, due to the hydration and dispersion effects, prominent vertical fractures formed. The interconnected fractures led to the chipping of the rock, rendering the structure unstable.

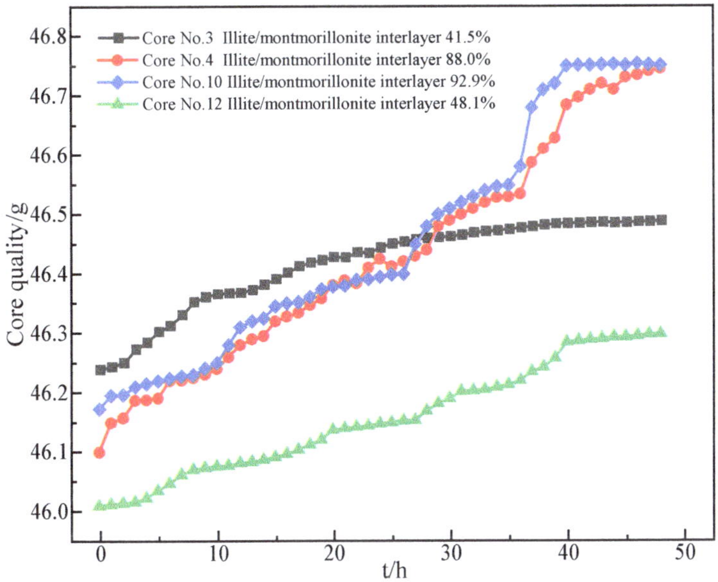

Figure 8. Results of water absorption experiments on rocks in loss zones.

Using thermal rolling recovery experiments, we further analyzed the thermal rolling recovery rates of these four types of rock particles under different conditions. The specific steps of the experiment were as follows: (1) Crush the rock using a core crusher to obtain rock particles. (2) Collect 1 kg of rock powder with a particle size of 0.149 mm using a vibrating sieve and store it in a wide-mouth bottle for later use. (3) Weigh 50.0 g of the prepared core particles and place them into a high-temperature vessel containing 350 mL of fluid. (4) Place the high-temperature vessel with the rock particles into a roller furnace and roll it at a formation temperature of 80 °C for 24 h. (5) Remove the high-temperature vessel and allow it to cool to room temperature. The liquid and rock samples inside the vessel were poured onto a sample sieve with a mesh size of 0.42 mm and rinsed in a basin of distilled water for 1 min. The remaining rock particles were then placed in a constant temperature drying oven for 4 h. After cooling, they were left to stand in air for 24 h before weighing, with the final weight recorded as m. The formula for calculating the thermal rolling recovery rate is $m/50 \times 100\%$. Figure 11 illustrates the variation of thermal rolling recovery rates for the different core samples at various concentrations of the inhibitor (modified natural polymer IND30). As shown in Figure 11, rocks 3 and 12, which had a relatively low illite/smectite interlayer content, exhibited generally high thermal rolling recovery rates (over 96%), indicating a weak overall hydration–dispersion capacity. Conversely, rocks 4 and 10, which had a high illite/smectite interlayer content, demonstrated thermal rolling recovery rates below 85%, indicating severe hydration–dispersion and a significant loss of rock particles during the thermal rolling process. However, with the addition of field inhibitors, the hydration–dispersion capacity of rocks 4 and 10 could be effectively reduced, leading to improved thermal rolling recovery rates. When the inhibitor concentration was 0.25%, the thermal rolling recovery rates for rocks 4 and 10 could be increased to over 94%.

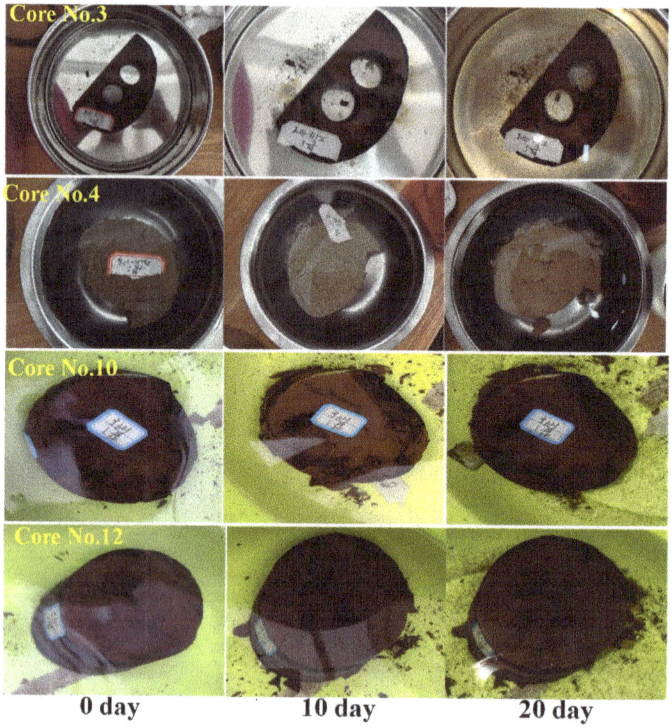

Figure 9. Hydration and dispersion behavior of reservoir rocks in loss zones at different soaking times.

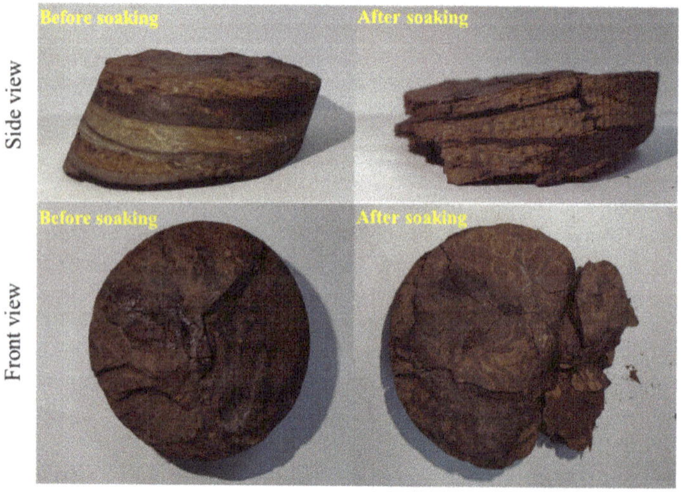

Figure 10. Comparison of main and side views of core 10 before and after soaking in distilled water.

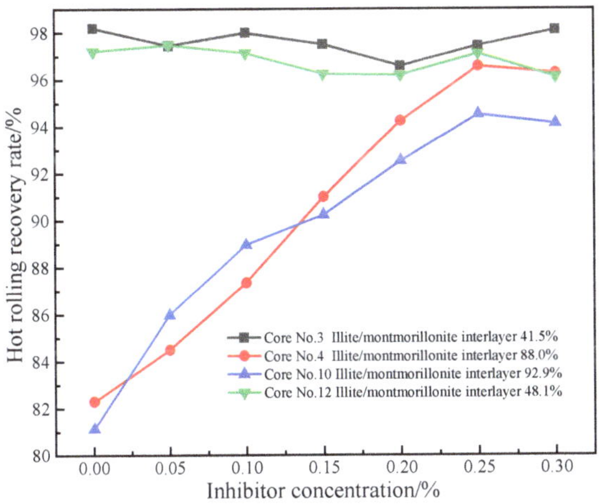

Figure 11. Variation of thermal rolling recovery rates of leakage layer rocks under different inhibitor concentrations.

4. Analysis of the Mechanical Properties of Rocks Surrounding the Wellbore

The content of Sections 1 and 2 has confirmed that the drilling loss in the Zhengning oilfield is related to natural fractures and the hydration–dispersion of the rocks. During the drilling process, the mechanical properties of the rocks surrounding the wellbore directly influence the loss pressure and the propagation of natural fractures. Based on this, we conducted true triaxial compression experiments on core samples from the same leakage layer to study the impact of natural fractures on the mechanical properties of leakage layer rocks. The experimental procedure was as follows: (1) Obtain standard cylindrical rock samples using a wire-cutting mechanism. (2) After polishing the surface of the rock samples, place them in the sample-holding device. (3) Apply confining pressure to the sample based on the in situ stress values obtained from well logging interpretation data. (4) Apply axial pressure and monitor the stress–strain changes of the sample.

As shown in Figure 12, after the triaxial compression experiments, the crack propagation in core samples 2 and 8 is illustrated, with red lines representing existing natural fractures and blue lines indicating fractures generated during the compression tests. Figure 13 presents the stress–strain curves obtained from these experiments. From Figure 12, it is evident that under identical pore-throat structures and mineral compositions, natural fractures significantly affected the mechanical properties of the leakage layer rocks. Cores without natural fractures primarily exhibited longitudinal fractures after triaxial compression, with a simple morphology and fewer occurrences. In contrast, cores with natural fractures exhibited complex fracture patterns, generating shear fractures that communicate with the natural fractures, leading to a substantial increase in the number of fractures upon failure. Combining insights from Figure 13, it is clear that the presence of natural fractures considerably reduced the compressive strength of the rocks under the same pore-throat structure and mineral composition. The rocks without natural fractures showed a steep stress–strain curve, indicating dense, hard rocks with strong elastic properties. Conversely, for the rocks with natural fractures, the curve was more gradual, suggesting higher Poisson's ratios and greater plasticity. Furthermore, considering the clay mineral content analysis from Section 1, the rocks with a higher kaolinite content (core sample 8) exhibited greater compressive strength compared with those with a lower kaolinite content (core sample 2).

Overall, this analysis indicates that the presence of natural fractures in leakage layer rocks significantly decreases their strength and increases the risk of wellbore collapse and failure.

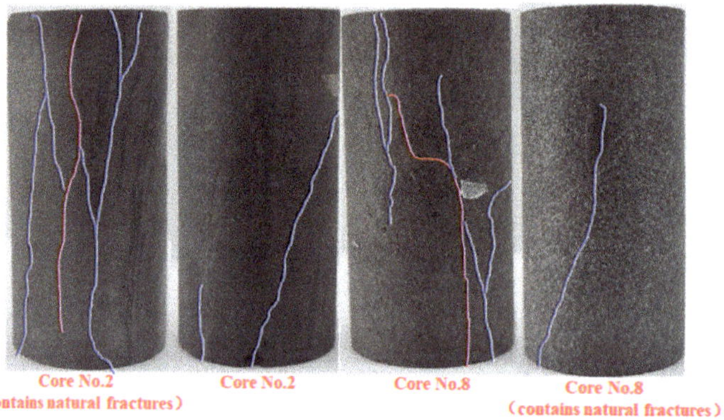

Figure 12. Results of true triaxial compression experiments for leakage layer core samples 2 and 8 (red lines represent natural fractures, while blue lines represent cracks generated during the mechanical experiments).

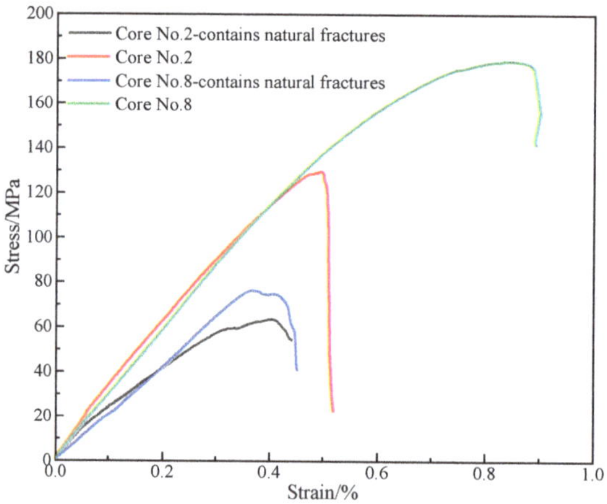

Figure 13. Stress–strain curves from true triaxial compression experiments for leakage layer core samples 2 and 8.

During the drilling process, the immersion of drilling fluid may cause hydration–dispersion in rocks, affecting their mechanical properties. In the previous two sections, we confirmed that when the relative content of illite/smectite interlayers exceeds 80% in clay minerals, the hydration–dispersion of the rocks becomes severe. For this section, we conducted a comparative analysis of the effects of drilling fluid immersion on the mechanical properties of core samples 4 and 8, selecting rocks without natural fractures for true triaxial mechanical experiments to eliminate the influence of natural fractures on the laboratory results. Figure 14 presents the results of the true triaxial compression experiments for leakage layer core samples 4 and 8, with the corresponding stress–strain curves shown in Figure 15. From Figure 14, it can be observed that in rocks that were not immersed in drilling fluid, the fracture morphology after failure was relatively simple, with fewer fractures. However, after immersion in drilling fluid, the fracture patterns became significantly more complex, with numerous branching fractures surrounding the main fracture, resulting in a substantial increase in the number of fractures. Figure 15 illustrates that under the same pore-throat structure and mineral composition, the compressive strength of the rocks decreased by over 25% following immersion in drilling fluid, indicating a significant impact of hydration–dispersion on the mechanical properties of the rocks. The stress–strain curve for rocks not immersed in drilling fluid was steep, indicating dense and hard rocks with a low Poisson's ratio. In contrast, after immersion in drilling fluid, the stress–strain curve shifted to the right, becoming more gradual, suggesting that hydration–dispersion increased the plasticity of the rocks and raised the Poisson's ratio. In summary, it can be concluded that during the drilling process in the Zhengning oilfield, the immersion of drilling fluid may be one of the reasons for the reduction in rock strength and the occurrence of wellbore loss incidents. For layers with a high content of illite/smectite interlayers, it is essential to add inhibitors during drilling to prevent the hydration–dispersion of the rocks.

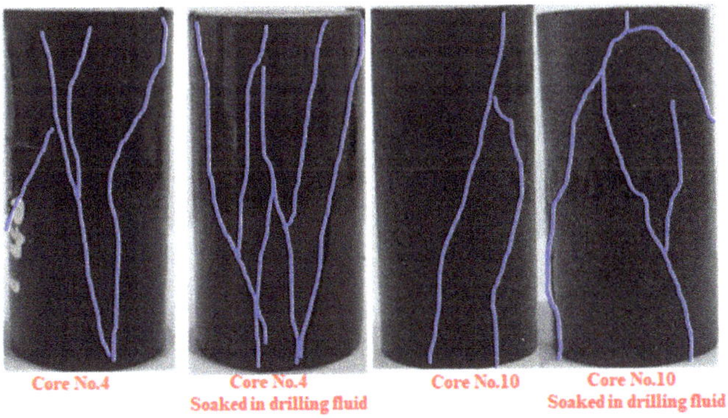

Figure 14. Results of true triaxial compression experiments for leakage layer core samples 4 and 8 (blue lines represent cracks generated during the mechanical experiments, with an immersion time of 48 h).

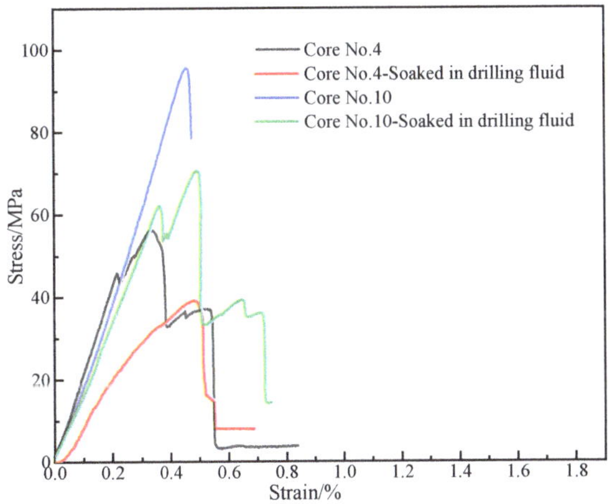

Figure 15. Stress–strain curves from true triaxial compression experiments for leakage layer core samples 4 and 10.

5. Conclusions

This study, based on core drilling from the Zhengning oilfield in the Ordos Basin, conducted laboratory experiments to investigate the mechanisms behind drilling loss incidents in the region. The main findings are as follows:

(1) The average porosity of the leakage layer reservoir rocks is 3.89%, with an average permeability of 0.044 mD. The clay mineral content exceeds 40%, generally high in the rocks.
(2) The leakage layer reservoir rocks are widely distributed with natural fractures. The macro natural fractures are most prevalent in the 0.2 mm to 0.3 mm range, accounting for over 30%, while micro natural fractures are primarily between 0.5 μm and 2 μm.
(3) When the relative content of illite/smectite interstratification in clay minerals exceeds 80%, the rock absorbs water rapidly, exhibits strong hydration–dispersion behavior, and presents a thermal recovery rate of less than 85%. After immersion in distilled water, the rock tends to flake off in chunks and the fracture width of the horizontal stratification fractures significantly increases, forming distinct longitudinal fractures.
(4) The presence of natural fractures significantly weakens the strength of the leakage layer rocks, causing them to exhibit plastic behavior. The strong hydration–dispersion destabilizes the rock structure, facilitating fracture interconnection and further reducing rock strength.
(5) During drilling, natural fractures provide pathways for drilling fluid loss. The macro natural fractures extend under borehole pressure, interconnecting with micro fractures and ultimately leading to increased fluid loss due to the hydration–dispersion effects.
(6) To improve drilling safety in the Zhengning oilfield, it is recommended to enhance the drilling fluid's sealing capacity for natural fractures of different fracture width levels and to strengthen the suppression effect of the drilling fluid on rock hydration–dispersion.

Author Contributions: Conceptualization, T.G. and S.Y.; methodology, Y.X.; software, L.Z.; validation, F.P. and X.S.; formal analysis, T.F.; investigation, H.W.; resources, D.Z.; data curation, Y.X.; writing—original draft preparation, L.Z.; writing—review and editing, X.S.; visualization, F.P.; supervision, T.F.; project administration, H.W.; funding acquisition, D.Z. All authors have read and agreed to the published version of the manuscript.

Funding: This research was funded by the National Natural Science Foundation of China, grant number U23B2089, and the Shaanxi Provincial Natural Science Basic Research Program Project, grant number 2024JC-YBQN-0554.

Data Availability Statement: The data presented in this study are available upon request from the corresponding author: 24111010032@stumail.xsyu.edu.cn.

Conflicts of Interest: Authors Tuan Gu, Shugang Yang, Yunfeng Xiao, Linpeng Zhang and Tao Fan are employed by Exploration and Development Research Institute of Liaohe Oilfield Company, PetroChina. The paper reflects the views of the scientists and not the company.

References

1. Yang, H.; Fu, J.H.; Wei, X.S.; Liu, X.S. Sulige field in the Ordos Basin: Geological setting, field discovery and tight gas reservoirs. *Mar. Pet. Geol.* **2008**, *25*, 387–400. [CrossRef]
2. Yao, J.L.; Deng, X.Q.; Zhao, Y.D.; Han, T.Y.; Chu, M.J.; Pang, J.L. Characteristics of tight oil in Triassic Yanchang Formation, Ordos Basin. *Pet. Explor. Dev.* **2013**, *40*, 161–169. [CrossRef]
3. Wang, H.; Zhou, D.; Zou, Y.; Zheng, P. Effect mechanism of seepage force on the hydraulic fracture propagation. *Int. J. Coal Sci. Technol.* **2024**, *11*, 43. [CrossRef]
4. Xu, C.; Kang, Y.; Chen, F.; You, Z. Fracture plugging optimization for drill-in fluid loss control and formation damage prevention in fractured tight reservoir. *J. Nat. Gas Sci. Eng.* **2016**, *35*, 1216–1227. [CrossRef]
5. Zhu, J.; You, L.; Li, J.; Kang, Y.; Zhang, J.; Zhang, D.; Huang, C. Damage evaluation on oil-based drill-in fluids for ultra-deep fractured tight sandstone gas reservoirs. *Nat. Gas Ind. B* **2017**, *4*, 249–255. [CrossRef]
6. Aljubran, M.; Ramasamy, J.; Albassam, M.; Magana-Mora, A. Deep Learning and Time-Series Analysis for the Early Detection of Lost Circulation Incidents During Drilling Operations. *IEEE Access* **2021**, *9*, 76833–76846. [CrossRef]
7. Lavrov, A. *Lost Circulation: Mechanisms and Solutions*; Gulf Professional Publishing: Oxford, UK, 2016.

8. Feng, F.; Gray, K.E. Modeling Lost Circulation Through Drilling-Induced Fractures. *SPE J.* **2018**, *23*, 205–223. [CrossRef]
9. Fuh, G.-F.; Morita, N.; Boyd, P.A.; McGoffin, S.J. A New Approach to Preventing Lost Circulation While Drilling. In Proceedings of the SPE Annual Technical Conference and Exhibition, Washington, DC, USA, 4–7 October 1992.
10. Howard, G.C.; Scott, P.P. An analysis and the control of lost circulation. *Trans. Am. Inst. Min. Metall. Eng.* **1951**, *192*, 171–182. [CrossRef]
11. Wang, H.; Zhou, D. Mechanistic study on the effect of seepage force on hydraulic fracture initiation. *Fatigue Fract. Eng. Mater. Struct.* **2024**, *47*, 1602–1619. [CrossRef]
12. Li, S.; Purdy, C. Maximum Horizontal Stress and Wellbore Stability While Drilling: Modeling and Case Study. In Proceedings of the SPE Latin American and Caribbean Petroleum Engineering Conference, Lima, Peru, 1–3 December 2010.
13. Ukaeru, F.C.; Igwilo, K.; Chikwe, A.; Ubanozie, O. Impact of Near-Wellbore Stress in Wellbore Integrity Analysis: Wellbore Fracture Strengthening Approach. *Eur. J. Sci. Innov. Technol.* **2024**, *4*, 187–206.
14. Jia, L.; Chen, M.; Hou, B.; Sun, Z.; Jin, Y. Drilling fluid loss model and loss dynamic behavior in fractured formations. *Pet. Explor. Dev. Online* **2014**, *41*, 105–112. [CrossRef]
15. Majidi, R.; Miska, S.Z.; Ahmed, R.; Yu, M.; Thompson, L.G. Radial flow of yield-power-law fluids: Numerical analysis, experimental study and the application for drilling fluid losses in fractured formations. *J. Pet. Sci. Eng.* **2010**, *70*, 334–343. [CrossRef]
16. Yang, J.; Sun, J.; Bai, Y.; Lv, K.; Zhang, G.; Li, Y. Status and Prospect of Drilling Fluid Loss and Lost Circulation Control Technology in Fractured Formation. *Gels* **2022**, *8*, 260. [CrossRef] [PubMed]
17. Zhiyu, W.; Changming, H.; Liang, L.; Chao, Y.; Hui, L. Theoretical Analysis of Drilling Fluid Flow for Maxi-Horizontal Directional Drilling. *J. Pipeline Syst. Eng. Pract.* **2024**, *15*, 04024050. [CrossRef]
18. Gazaniol, D.; Forsans, T.; Boisson, M.J.F.; Piau, J.-M. Wellbore Failure Mechanisms in Shales: Prediction and Prevention. *J. Pet. Technol.* **1995**, *47*, 589–595. [CrossRef]
19. Wang, B.; Sun, J.; Shen, F.; Li, W.; Zhang, W. Mechanism of wellbore instability in continental shale gas horizontal sections and its water-based drilling fluid countermeasures. *Nat. Gas Ind. B* **2020**, *7*, 680–688. [CrossRef]
20. Zhao, X.; Qiu, Z.; Wang, M.; Xu, J.; Huang, W. Experimental investigation of the effect of drilling fluid on wellbore stability in shallow unconsolidated formations in deep water. *J. Pet. Sci. Eng.* **2019**, *175*, 595–603. [CrossRef]
21. Ma, J.; Pang, S.; Zhang, Z.; Xia, B.; An, Y. Experimental Study on the Polymer/Graphene Oxide Composite as a Fluid Loss Agent for Water-Based Drilling Fluids. *ACS Omega* **2021**, *6*, 9750–9763. [CrossRef] [PubMed]
22. Nasser, J.; Jesil, A.; Mohiuddin, T.; Ruqeshi, M.A.; Devi, G.; Mohataram, S. Experimental Investigation of Drilling Fluid Performance as Nanoparticles. *World J. Nano Sci. Eng.* **2013**, *3*, 5. [CrossRef]
23. Al-Hameedi, A.T.T.; Alkinani, H.H.; Dunn-Norman, S.; Al-Alwani, M.A.; Alshammari, A.F.; Alkhamis, M.M.; Mutar, R.A.; Al-Bazzaz, W.H. Experimental investigation of environmentally friendly drilling fluid additives (mandarin peels powder) to substitute the conventional chemicals used in water-based drilling fluid. *J. Pet. Explor. Prod. Technol.* **2020**, *10*, 407–417. [CrossRef]
24. Ezeakacha, C.P.; Salehi, S. Experimental and statistical investigation of drilling fluid loss in porous media: Part 2 (Fractures). *J. Nat. Gas Sci. Eng.* **2019**, *65*, 257–266. [CrossRef]
25. Srivatsa, J.T.; Ziaja, M.B. An Experimental Investigation on Use of Nanoparticles as Fluid Loss Additives in a Surfactant—Polymer Based Drilling Fluids. In Proceedings of the International Petroleum Technology Conference, Bangkok, Thailand, 15–17 November 2011.
26. Kong, C.; Sun, Y.; Li, C.; Zhao, J.; Zhu, X. Numerical simulation study on impact factors to dynamic filtration loss. *J. Pet. Explor. Prod. Technol.* **2024**, *14*, 593–607. [CrossRef]
27. Yang, L.; Cheng, Z.G. Numerical Simulation of Large-Diameter Annular Pressure Loss in Riser Segment of Deep-Water Drilling. *Adv. Mater. Res.* **2013**, *868*, 510–516. [CrossRef]
28. Xu, C.; Liu, L.; Yang, Y.; Kang, Y.; You, Z. An innovative fracture plugging evaluation method for drill-in fluid loss control and formation damage prevention in deep fractured tight reservoirs. *Fuel* **2024**, *358*, 130123. [CrossRef]
29. ALi, A.; Gao, S.; Zhang, G.; Zeng, Y.; Hu, Y.; Zhai, R.; Dong, A.; Zhang, J. A Review in Polymers for Fluid Loss Control in Drilling Operations. *Macromol. Chem. Phys.* **2024**, *225*, 2300390. [CrossRef]
30. Sun, J.; Yang, J.; Bai, Y.; Lyu, K.; Liu, F. Research progress and development of deep and ultra-deep drilling fluid technology. *Pet. Explor. Dev. Online* **2024**, *51*, 1022–1034. [CrossRef]
31. Wu, S.; Li, L.; Li, X.; Yue, Z.Q. Characterizing strength and location of continental oil shale with drilling process monitoring in Southern Ordos Basin, China. *J. Rock Mech. Geotech. Eng.* **2024**. [CrossRef]
32. Liu, J.; Zhang, G.; Bai, J.; Ding, W.; Yang, H.; Liu, Y. Quantitative prediction of the drilling azimuth of horizontal wells in fractured tight sandstone based on reservoir geomechanics in the Ordos Basin, central China. *Mar. Pet. Geol.* **2022**, *136*, 105439. [CrossRef]

33. Zhang, X.; Dang, B.; Wang, X.; Luo, S.; Chen, B.; Zheng, L. Acid-Soluble Drilling Fluid in the Northern Carbonate Reservoir of the Yishan Slope in the Ordos Basin. *Energies* **2023**, *16*, 6020. [CrossRef]
34. Zhang, X.; You, L.; Kang, Y.; Zhang, C.; Zhang, G.; Tan, Q. Formation damage control of saline-lacustrine fractured tight oil reservoir during well drilling. *Arab. J. Geosci.* **2020**, *13*, 1087. [CrossRef]

minerals

MDPI

Article

Preparation and Performance Evaluation of Small-Molecule Ammonium as a Shale Hydration Inhibitor

Quande Wang [1,2], Huifeng He [3], Yarui Zhao [4], Jing Rui [4], Ruichen Jiang [4], Michal Slaný [5,6,*], Gang Chen [1,2] and Xuefan Gu [1,2,*]

1 Engineering Research Center of Oil and Gas Field Chemistry, Universities of Shaanxi Provence, Xi'an Shiyou University, Xi'an 710065, China; gangchen@xsyu.edu.cn (G.C.)
2 Shaanxi Province Key Laboratory of Environmental Pollution Control and Reservoir Protection Technology of Oilfields, Xi'an Shiyou University, Xi'an 710065, China
3 Changqing Drilling Company of CCDC, China National Petroleum Corporation, Xi'an 710060, China
4 No. 11 Oil Production Plant, Changqing Oilfield Company, Qinyang 745000, China
5 Institute of Inorganic Chemistry, Slovak Academy of Sciences, Dúbravská cesta 9, 845 36 Bratislava, Slovakia
6 Department of Materials Engineering and Physics, Faculty of Civil Engineering, Slovak University of Technology, Radlinského 11, 810 05 Bratislava, Slovakia
* Correspondence: michal.slany@savba.sk (M.S.); xuefangu@xsyu.edu.cn (X.G.)

Abstract: In this paper, small-molecule quaternary ammonium salts were synthesized by N-alkylation to inhibit hydration swelling and hydration dispersion. The prepared small-molecule quaternary ammonium salt was characterized by Fourier transform infrared (FTIR) spectroscopy, Thermogravimetric analysis (TGA), particle size analysis and Scanning electron microscopy (SEM), and its performance as an inhibitor in clay was evaluated by an anti-swelling test and a linear swelling test. The results show that small-molecule quaternary ammonium salt (TEE-2) synthesized by triethanolamine and epichlorohydrin in ethanol with a molar ratio of 1:1.5 can successfully inhibit the hydration swelling and dispersion of clay. The anti-swelling rate of TEE-2 was 84.94%, the linear swelling rate was 36.42%, and the linear swelling rate of 0.5% TEE-2 was only 29.34%. The hydration swelling of clay in 0.5% TEE-2 solution was significantly inhibited. The hydration inhibition mechanism of the small-molecule quaternary ammonium salt inhibitor 0.5% TEE-2 was analyzed by FTIR, SEM and TGA. It was considered that 0.5% TEE-2 has strong hydration inhibition, which was realized by infiltration and adsorption on the clay surface. Small-molecule quaternary ammonium salts were beneficial for maintaining wellbore stability and reducing the risk of wellbore instability.

Keywords: hydration; infiltration; adsorption; wellbore stability; bentonite

1. Introduction

Shale oil has emerged as one of the most outstanding technologies in the world in recent years [1]. In the process of oil field drilling, due to the hydration swelling of water-sensitive shale, drilling instability problems such as drilling scouring, pipe jamming, rock debris disintegration and bit balling often occur during shale formation [2,3]. According to the chemical characteristics of shale and drilling fluid, when water-sensitive shale (with high montmorillonite content) is immersed in water-based drilling fluid, shale may expand and disperse rapidly [4,5]. Therefore, many shale inhibitors have been widely used in water-based drilling fluids. Unfortunately, environmental requirements limit the use of most shale inhibitors [6–8].

As is well known, KCl is one of the inorganic salt inhibitors used to inhibit the hydration swelling of clay. K^+ can enter the clay layer and be adsorbed on the surface, expelling the interlayer water and reducing the interlayer spacing of the clay [9]. However, KCl has limited ability to inhibit clay hydration swelling and has an impact on the environment and rheological properties of drilling fluids. Polymer amine inhibitors have a large number of amine groups that adsorb onto clay surfaces, forming dense films to reduce the hydration of clay. However, the molecular structure of polymer amine inhibitors is too large to effectively embed between clay layers [10].

In recent years, small-molecule amines have attracted extensive attention from researchers because of their good inhibition, lubricity and stable rheology. Therefore, we have proposed the creation of environmentally friendly small-molecule inhibitors with low molecular weight. Small-molecule amine inhibitors can provide multiple adsorption sites on the surface of clay and enhance the adsorption of inhibitors in the clay [11,12]. The binding of clay is mainly realized through hydrogen bonding, anchoring, electrostatic adsorption and hydrophobic action, which effectively inhibits the hydration, swelling and dispersion of clay [13,14]. Small-molecular amines have excellent compatibility with traditional additives and can meet the requirements of environmental protection. They have been applied to many water-based drilling fluids and have very broad application prospects.

In this paper, epichlorohydrin and triethanolamine were used to prepare small-molecule shale inhibitors. The performance was evaluated by experiments such as anti-swelling, linear swelling and drilling fluid performance evaluation. The mechanism of the inhibitor was discussed in detail in terms of its FTIR, SEM, TGA and zeta potential.

2. Experimental Materials and Methods

2.1. Materials and Reagents

Epichlorohydrin was purchased from Chengdu Kelong Chemical Reagent Factory. Triethanolamine and ethanol were purchased from Tianjin Fuyu Fine Chemical Co., Ltd., Tianjin, China. Industrial-grade triethylenetetramine was taken from the site of Changqing Oilfield. Calcium- and sodium-based bentonite were purchased from Xi'an Fengyun Chemical Co., Ltd., China. Sodium carbonate was purchased from Tianjin Shengao Chemical Reagent Co., Ltd., China. Potassium chloride was purchased from Tianjin Zhiyuan chemical reagent factory. Heteropolysaccharide (KD-03), rubber powder (RP), carboxymethyl cellulose (CMC) and modified starch (MS) were purchased from Yangzhou Runda Oilfield Chemical Co., Ltd., Yangzhou, China.

Chemical Composition of Raw Bentonites

Bentonite is an absorbent swelling clay consisting mainly of montmorillonite (a type of smectite), which can be either calcium- or sodium-type. Na–montmorillonite has a significantly greater swelling capacity than Ca–montmorillonite. The common smectite of bentonite is montmorillonite-beidellite The main use of bentonite is in drilling mud and as an absorbent, binder, cleaner, and carrier of fertilizers and others. The oxide composition of calcium and sodium-based bentonite based on XRF is presented in Table 1.

Table 1. Chemical composition of raw bentonites.

	Oxide Composition/Mass%							
	SiO_2	Al_2O_3	CaO	Fe_2O_3	MgO	K_2O	Na_2O	LOI
Ca-bentonite	65.48	15.85	2.85	2.54	1.85	1.78	1.45	7.35
Na-bentonite	65.20	12.94	3.78	3.45	2.65	0.35	2.58	8.65

2.2. Synthesis and Nomenclature of Inhibitors

Triethanolamine and epichlorohydrin were added to the flask in the ratio stated in Table 2. Ethanol was used as the solvent and the reaction was refluxed for 4 h. The mixture was then cooled to room temperature and the ethanol was evaporated to create the final product. The reaction mechanism is shown in Figure 1.

Table 2. Nomenclature of inhibitors.

Reagent	Reagent	Solvent	Proportion	Nomenclature
Triethanolamine	Epichlorohydrin	Ethanol	1.0:1.0	TEE-1
			1.0:1.5	TEE-2
			1.0:3.0	TEE-3
		Distilled water	1.0:1.0	TEW-1
			1.0:1.5	TEW-2
			1.0:3.0	TEW-3
		Glycerol	1.0:1.0	TEG-1
			1.0:1.5	TEG-2
			1.0:3.0	TEG-3

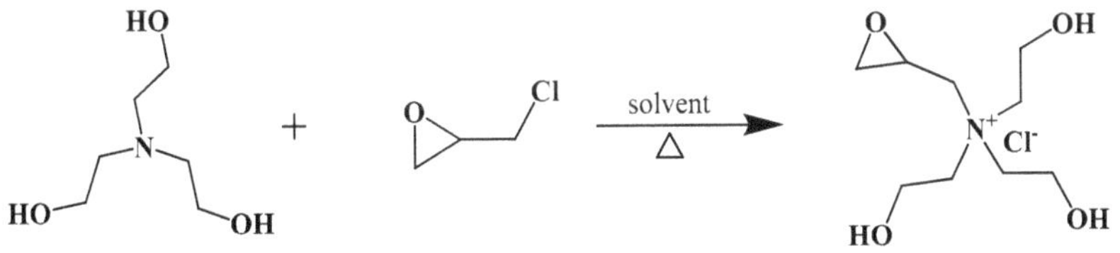

Figure 1. Synthesis mechanism of inhibitors.

2.3. Optimization of Synthetic Inhibitors

The inhibition performance of synthetic products on clay is affected by the material ratio, concentration and reaction medium of synthetic reaction, and there are many synthetic inhibitors. Therefore, the synthetic products with the best inhibition performance were preliminarily selected through inhibition performance parameters such as the anti-swelling rate and linear swelling rate. Subsequently, we evaluated the inhibition efficiency in water-based drilling fluid and explored its mechanism.

2.4. Anti-Swelling Rate

The industry-standard SY/T 5971-1994 evaluation method of clay stabilizer for drilling fluid was referred to in order to evaluate the influence of the inhibitor on the anti-swelling rate of bentonite. Inhibitor solutions of different concentrations were prepared. First, 0.5 g bentonite was weighed and added to a 10 mL centrifuge tube. A certain amount of inhibitor solution was added to the centrifuge tube and then fully stirred and shaken. After standing for 2 h, it was centrifuged at the speed of 1500 r/min for 15 min, and the volume was recorded (V_a). The inhibitor solution was replaced with water and kerosene, respectively, and the swelling volume of bentonite in water and kerosene was recorded as V_b and V_0, respectively. The calculation formula for the anti-swelling rate of bentonite is shown in Equation (1):

$$\mathrm{B} = \frac{V_b - V_a}{V_b - V_0} \times 100\% \quad (1)$$

where B is the anti-swelling rate of bentonite. V_a is the swelling volume of bentonite in inhibitor solution in mL. V_b is the swelling volume of bentonite in water, also in mL. V_0 is the swelling volume of bentonite in kerosene in mL.

2.5. Linear Swelling

The industry-standard SY/T 6335-1997 shale inhibitor evaluation method for drilling fluid was referred to for evaluating the influence of inhibitor on the linear swelling rate of bentonite. The calculation formula of linear swelling rate of bentonite is shown in Equation (2):

$$\mathrm{Sr} = \frac{Ro}{\Delta L} \times 100\% \quad (2)$$

where Sr is linear swelling rate of bentonite. *Ro* is the swelling of bentonite in mm. ΔL is the core thickness in mm.

2.6. Drilling Fluid Evaluation

The 4% calcium clay-based mud preparation was conducted as follows: Calcium bentonite (14 g) and sodium carbonate (0.7 g) were added to tap water (350 mL), stirred at high speed for 2 h and aged at 298 K for 24 h before use [15]. To prepare the treatment-based mud and treatment agent, they were aged for 6 h, stirred at high speed for 10 min, and their performances were tested [16]. The rheological properties, filtration properties, and lubrication properties of drilling fluid, such as the AV (apparent viscosity), PV (plastic viscosity), YP (yield point), FL (API filtration) and t g (friction coefficient) were evaluated. A viscometer (ZNN-D6S, Hetongda Co. Ltd. Qingdao, China), medium-pressure filtration instrument (GJSS-B12K, Haitongda Co. Ltd. Qingdao, China), and viscosity coefficient instrument (Qingdao Hetongda Co. Ltd. Qingdao) were used according to the formulas described in the Chinese National Standard GB/T 16783.1-2006.

2.7. Inhibition Mechanism Study

2.7.1. FTIR Analysis

The dried inhibitor samples were ground. During the test, the ground samples were mixed with KBr at a ratio of 1:200, put into the tablet press and pressed into transparent pellets, and the soil samples were scanned and analyzed by an infrared spectrometer (Nicolet 6700) [17]. The FTIR spectra were measured using a resolution of 4 cm^{-1}, 64 scans, and a measurement region of 4000–400 cm^{-1}.

2.7.2. Particle Distribution Measurement

The dried inhibitor samples were used to measure the particle size using a laser particle size experiment so as to obtain the median particle size and average particle size of bentonite particles in mud treated with treatment agent. The change in bentonite particle size was analyzed according to these data [18].

2.7.3. Zeta Potential Measurement

The zeta potential of the supernatant of the solution was measured on an omni multiangle particle size and high-sensitivity zeta potential analyzer. The changes in the zeta potential of graphite with different dosages of adsorbent were analyzed [19].

2.7.4. SEM Analysis

The bentonite samples were dispersed in the inhibitor solution and hydrated for 24 h, then the water was separated from the solution and dried at 105 °C for SEM. The surface morphology of the bentonite samples was evaluated by a digital microscope imaging scanning electron microscope (model Vega 3, Tescan, serial no. HI-2102-0003) at a 40.0 kV accelerating voltage on the basis of the reported method [20]. Prior to SEM measurements, the samples were coated with gold.

2.7.5. TGA

The bentonite samples were dispersed in the inhibitor solution and hydrated for 24 h, then the water was separated from it and dried at 105 °C for TGA and SEM. The TGA experiment was conducted on a TGA/DSC thermal analysis instrument (1/1600, METTLER TOLEDO, Inc., Columbus, OH, USA) at a ramp of 20 °C/min from room temperature to 825 °C under nitrogen flow.

2.7.6. XRF Analysis

The calcium and sodium bentonite samples were characterized by X-ray Fluorescence (XRF) and clay samples were obtained through Phillips' XRF PW2400 model (Bruker,

Billerica, Massachusetts, USA), with the scintillation detector having a voltage 40 mV, current of 40 mA and rhodium anode X-ray tube.

3. Results and Discussion

3.1. Optimization Results of Synthetic Inhibitors

The control variable method was adopted, and the optimal solvent, reaction molar ratio and dosage were screened through anti-swelling and linear swelling experiments. The inhibitors with strong inhibitory performance have been identified.

3.1.1. Anti-Swelling Rate

The inhibition performance of the product was preliminarily explored, and the anti-swelling performance of the synthetic product on bentonite was evaluated. The anti-swelling test results are shown in Figure 2. TEE, TEW and TEG have certain anti-swelling effects on bentonite, and their anti-swelling rate is related to the ratio of raw materials, solvent, and inhibitor concentration. It can be seen from Figure 2 that the inhibitory effect of TEE series was significantly better than that of TEW and TEG series. The results show that when the reaction medium of epichlorohydrin and triethanolamine was ethanol, the inhibition performance of the synthetic product was better. It can be seen from Figure 2a that when ethanol was used as the solvent, the inhibitory effect of TEE-2 was better than that of TEE-1 and TEE-3. The results show that when the molar ratio of triethanolamine to epichlorohydrin was 1:1.5, the synthesized product achieved a better inhibition performance. It can be seen from Figure 2a that when ethanol was used as the solvent and the molar ratio was 1:1.5, the anti-swelling rate increased with the increase in inhibitor concentration. When TEE-2 was added in a concentration of 0.5%, the maximum anti-swelling rate was 84.94%.

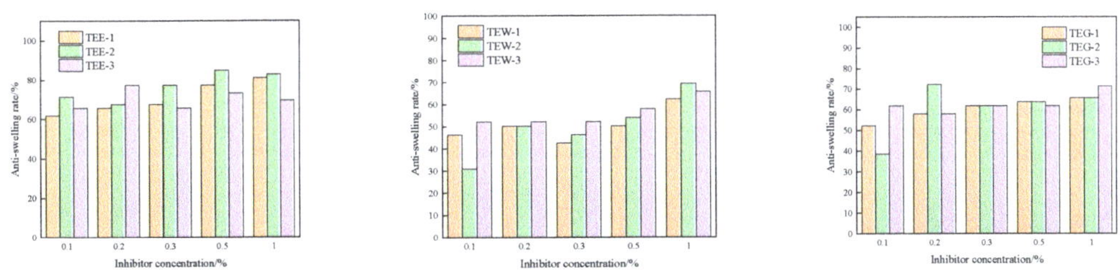

(**a**) TEE (**b**) TEW (**c**) TEG

Figure 2. The effect of inhibitor concentration on anti-swelling rate.

3.1.2. Linear Swelling

During the drilling process, wellbore collapse and instability have a very bad and irreversible impact on the exploitation of the oilfield [21]. The linear swelling rate measured in laboratory experiments can reflect the degree of wellbore collapse and instability to a certain extent. Therefore, the inhibition performance of synthetic inhibitors on bentonite can be further explored by linear swelling experiments. The experimental results are shown in Figure 3. It can be seen from Figure 3 that at a certain concentration and molar ratio, ethanol as a solvent synthesized inhibitor had a lower swelling rate in bentonite. These concentrations were 40.00%, 36.42% and 41.43%, respectively. It can be seen from Figure 4 that when ethanol was used as solvent and at a certain concentration, the molar ratio of TEE inhibitor was 1:1.5, indicating that it has a better inhibitory effect. The linear swelling rate of 0.1% TEE-2 was 36.42%. Compared with the inhibition of tap water and 4% KCI, the swelling rate decreased by 25.89% and 8.73%. It can be seen from Figure 5 that when ethanol was used as solvent and the molar ratio was 1:1.5, the inhibition effect of 0.5% TEE-2 was the best. The linear swelling rate of 0.5% TEE-2 was only 29.34%. It is obvious that the inhibitory effect of 0.5% TEE-2 was better than that of other concentrations of TEE-2.

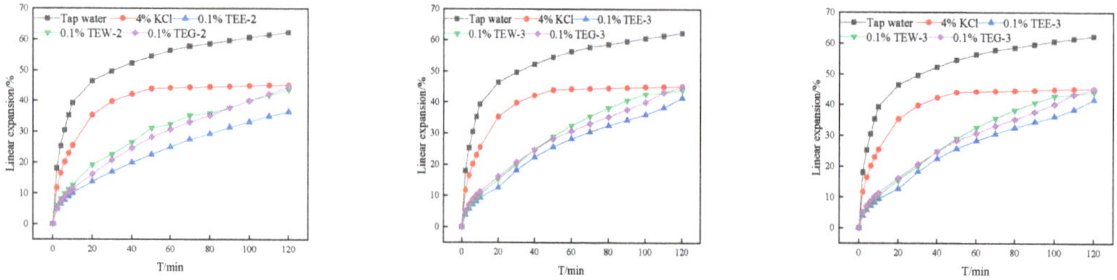

(**a**) The molar ratio 1:1. (**b**) The molar ratio 1:1.5. (**c**) The molar ratio 1:2.

Figure 3. The effect of different solvents on linear swelling of bentonite.

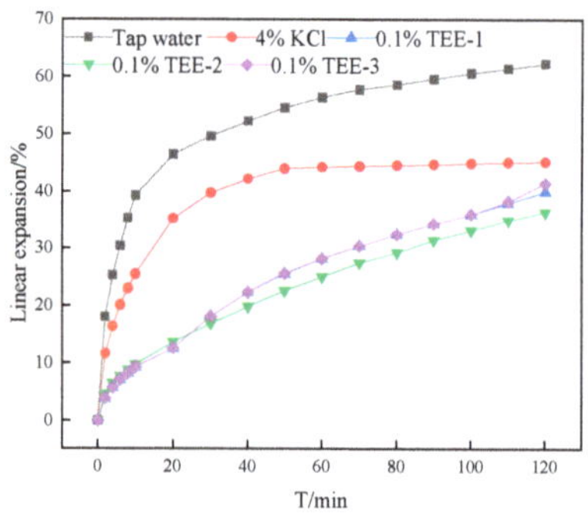

Figure 4. The effect of TEE with different molar ratios on linear swelling of bentonite.

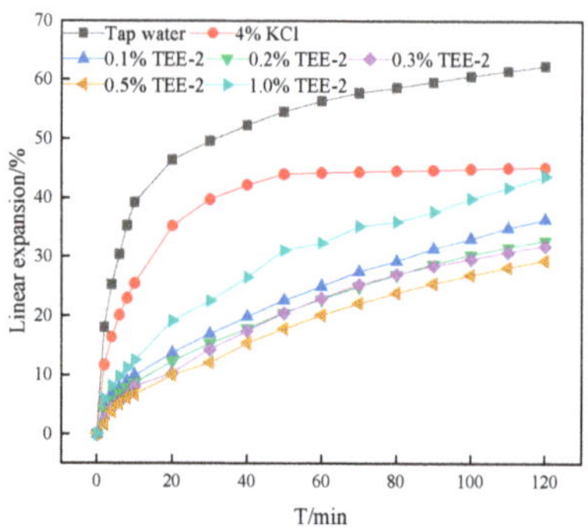

Figure 5. The effects of different concentrations of TEE-2 on linear swelling of bentonite.

3.2. Drilling Fluid Evaluation

A certain amount of TEE-2 was added to the treated mud of low-viscosity CMC, RP, KD-03, and MS. The rheological parameters of several types of water-based drilling fluids were effectively improved, the filtration loss was reduced, and the suspension force of drilling fluid was enhanced. TEE-2 has good compatibility with CMC, RP, MS, and KD-03, and its compatibility with low-viscosity CMC was the best. We added 0.5% TEE-2 to low-viscosity CMC, RP, MS, and KD-03-treated mud at room temperature. The performance parameters of drilling fluid are shown in Table 3. It can be seen from Table 3 that the rheological parameters of drilling fluid increase after the addition of treatment agent. After 0.5% TEE-2 was added to CMC-treated mud, the AV decreased by 5.00%, the PV increased by 22.22%, the YP increased by 2.50 times, and the filtration loss and slider resistance coefficient also increased to a certain extent. After 0.5% TEE-2 was added to the RP-treated mud, the AV, PV, and YP of the RP-treated mud increased. With the increase in AV, the suspension capacity was enhanced, and the resistance coefficient of the slider also increased to a certain extent, which had the effect of increasing viscosity. KD-03 has a lubricating effect in water-based drilling fluid and leads to filtration reduction. After 0.5% TEE-2 was added to KD-03-treated mud, AV and PV achieved 1.07 and 1.25 times the rheological parameters of KD-03-treated mud, respectively, and the filtration loss and the resistance coefficient of sliding block were basically unchanged. MS treatment agent can effectively control the filtration of drilling fluid, adjust the rheology of drilling fluid, and have a good anti-sloughing effect. After 0.5% TEE-2 is added to the MS-treated mud, the filtration rate is further effectively controlled, and the filtration rate is reduced from 10.8 mL to 8.5 mL. The AV, PV and YP/PV are 1.09, 1.67 and 1.67 times those of modified starch data, respectively.

Table 3. Effect of 0.5% TEE-2 on compatibility of water-based drilling fluid.

Treatment Agent	AV /(mPa•s)	PV /(mPa•s)	YP /Pa	YP/PV /(m•s)	FL/mL	tg
Base mud	2.0	1.4	0.6	0.43	15.9	0.0437
Base mud + 0.5% TEE-2	3.0	2.0	1.0	0.50	17.0	0.1673
0.5%CMC	10.0	9.0	1.0	0.11	4.8	0.0875
0.5% CMC + 0.5% TEE-2	9.5	7.0	2.5	0.36	4.5	0.1853
1.0% RP	17.5	15.1	2.4	0.16	2.0	0.1495
1.0% RP + 0.5% TEE-2	10.4	8.6	1.8	0.21	2.2	0.1228
0.3%KD-03	2.8	2.0	0.8	0.38	14.0	0.0437
0.3% KD-03 + 0.5% TEE-2	3.0	2.5	0.5	0.20	15.5	0.0963
1.0%MS	6.9	6.0	0.9	0.15	10.8	0.1405
1.0% MS + 0.5% TEE-2	7.5	6.0	1.5	0.25	8.5	0.1673

3.3. FTIR Analysis

FTIR was used to analyze the functional groups of bentonite and inhibitor molecules, as shown in Figure 6. It can be seen from FTIR spectra (Figure 6) that bentonite treated with water contains the characteristic band near 3620 cm^{-1} and is assigned to the stretching vibration of structural OH groups. The characteristic band at 3422 cm^{-1} is related to the stretching vibration of O-H groups of water in montmorillonite and hygroscopic KBr. The characteristic band at 3268 cm^{-1} observed in the IR spectrum of bentonite treated with TEE-2 is the absorption vibration band of ammonium salt molecules (vNH). This finding confirms the correct preparation of organo-bentonite samples. The strong band at 1630 cm^{-1} is assigned to the bending vibration of OH groups. The characteristic band near 1034 cm^{-1} is related to the vSi-O. The absorption band near 469 cm^{-1} is caused by the coupling vibration of Si-O-Si [22]. The main functional groups of quaternary ammonium salt inhibitors are adsorbed on the surface of bentonite, further restricting the swelling and hydration of shale.

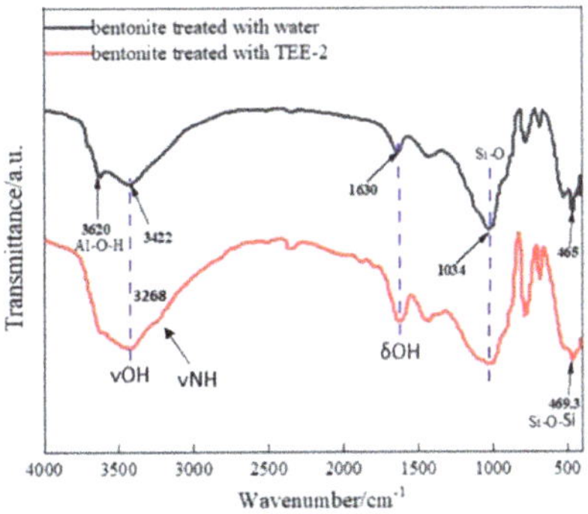

Figure 6. FTIR spectra of the bentonite treated with water and TEE-2.

3.4. Particle Distribution Measurement

Inhibitors have a certain microscopic effect on the particle size of bentonite. The changes in particle size of untreated bentonite particles and bentonite particles treated with different solutions were explored by laser particle-size analysis (Table 4), and the influence of synthetic products on the particle size of bentonite was analyzed [23]. The changes in the particle size of sodium bentonite after 0.5% TEE-2 was added before hydration (BH) and after hydration (AH) were investigated. The results of particle size distribution are shown in Figure 7, and the average particle size and median particle size are shown in Table 3. It can be seen from Table 3 that the average particle size and median particle size of BH were 14.270 μm and 11.020 μm, respectively. The average particle size and median particle size of hydrated bentonite in clean water are 7.903 μm and 4.660 μm, respectively. After the addition of 0.5% TEE-2, the average particle size without hydration was reduced to 47.13% of the original particle size, and the median particle size was reduced to 26.06% of the original median particle size. After hydration, the average particle size and median particle size increased slightly, growing to 1.04 times and 1.45 times those of the original hydration blank group, respectively. In short, the median particle size and average particle size became smaller regardless of hydration status. It can be seen from Figure 7 that the addition of 0.5% TEE-2 to bentonite can significantly inhibit the hydration and dispersion of bentonite.

Table 4. Average particle size and median particle size of sodium bentonite in 0.5% TEE-2 solution.

Treatment of Bentonite	The Average Particle Size/μm	Median Particle Size/μm
BH	14.270	11.020
AH	7.903	4.660
BH(0.5% TEE-2)	7.544	8.148
AH(0.5% TEE-2)	8.185	6.761

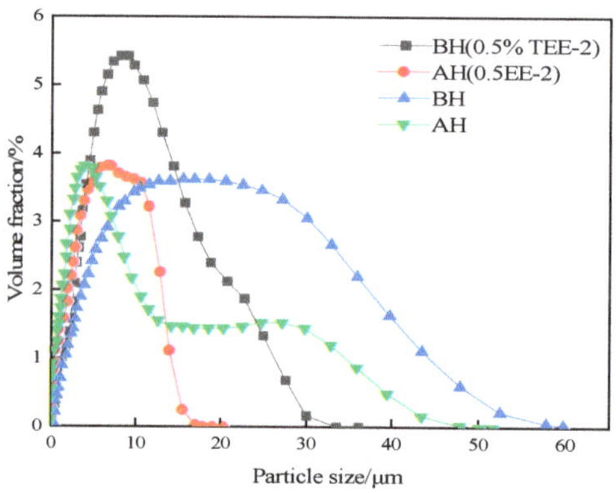

Figure 7. Effect of 0.5% TEE-2 on particle size distribution of sodium bentonite BH and AH.

3.5. Zeta Potential Measurement

The hydration swelling dispersion of bentonite is caused by many factors, which depend not only on the composition and structure of bentonite, but also on the composition of exchangeable cations and the properties of dispersion medium [24]. The zeta potential value of dispersion medium solution is closely related to the dispersion state of bentonite. Therefore, the inhibition performance of 0.5% TEE-2 on the hydration swelling of bentonite was explored by detecting the zeta potential of the adsorbed electric double layer on the surface of bentonite in different concentrations of 0.5% TEE-2 solution. The zeta potential of 0.5% TEE-2 solution is shown in Figure 8. Negative charge was formed by isomorphic substitution of octahedron. The increase in positive charge in the interlayer space breaks the charge balance. The zeta potential of bentonite treated with clean water was −21.41 mv, which has good dispersion in water. With the addition of 0.5% TEE-2, the zeta potential of the solution decreases, the absolute value of zeta potential increases, and the system achieves a stable state again, which effectively inhibits the hydration and dispersion of bentonite.

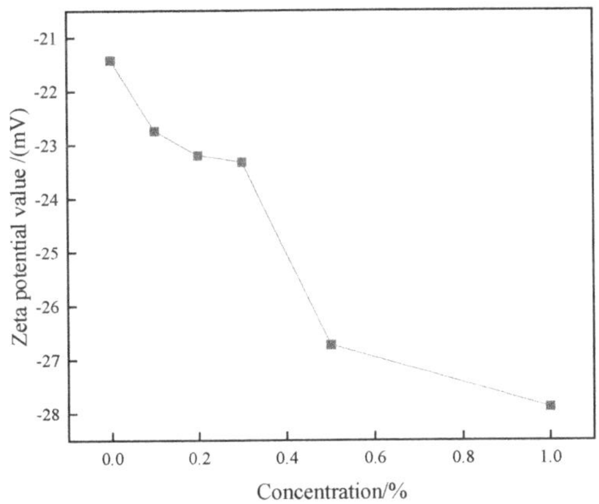

Figure 8. Effect of 0.5% TEE-2 concentration on zeta potential of electric double layer adsorbed on bentonite surface.

3.6. SEM Analysis

The micro-morphology of bentonite particles treated and dried with 0.5% TEE-2 and clean water was analyzed by SEM. The effect of 0.5% TEE-2 on the microstructure of bentonite is explored in Figure 9. Figure 9a shows the SEM images of the microstructure of unhydrated bentonite. Figure 9b,c depict the microstructure of bentonite after hydration treatment in clean water and 0.5% TEE-2 solution for 24 h. It can be seen from the microstructure diagram that bentonite was strongly dispersed in clean water. After the addition of 0.5% TEE-2, the fine particles of bentonite decreased significantly. This shows that after the addition of 0.5% TEE-2, the inhibitor enters the bentonite layers and binds the bentonite layers together through electrostatic adsorption and hydrogen bonding. Thus, it can effectively inhibit the hydration swelling and dispersion of bentonite. Macroscopically, the large particles of bentonite increase in number and the particle size decreases, showing a certain inhibition.

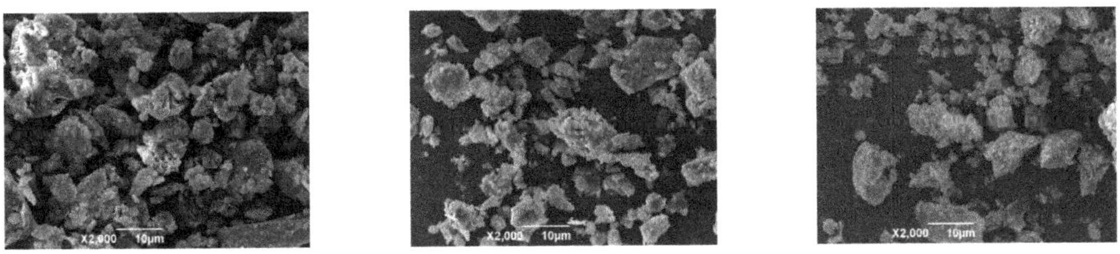

(**a**) Un-hydrated bentonite (**b**) Hydrated bentonite (**c**) 0.5% TEE-2

Figure 9. SEM image of bentonite under different conditions.

3.7. TGA

During the heating process, with the increase in temperature, the adsorbed water, interlayer water and hydroxyl water in bentonite are removed in turn [25]. Therefore, the weight loss rate of bentonite with inhibitors can be determined by TGA. The effect of inhibitors was evaluated based on the hydration swelling and dispersion of bentonite, as shown in Figure 10. It can be seen from Figure 10 that as the temperature gradually increased, the water between the clay layers began to evaporate, and the weight of the clay particles decreased. And the weight loss rate of bentonite processed in TEE-2 was 10%. However, the weight loss rate of untreated bentonite was 13%. This result indicates that TEE-2 can significantly prevent water molecules from penetrating shale and has a certain inhibitory effect on the hydration swelling and dispersion of bentonite.

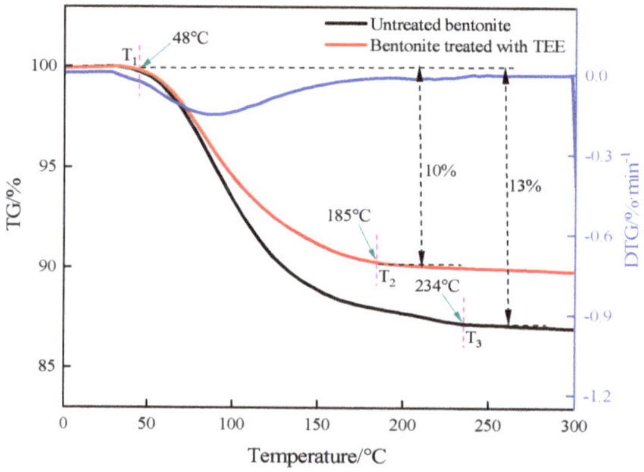

Figure 10. TGA curve of inhibitor TEE-2.

3.8. Mechanism

The mechanism of inhibiting hydration and swelling of quaternary ammonium salt small-molecule inhibitors in clay was systematically discussed based on N-alkylation and the diffusion double-layer theory. It is generally believed that the core action of inhibitors is to reduce the repulsion between clay crystal layers and prevent contact between water and clay particles [26,27]. It can be seen from Figure 11 that under the action of potential difference and water potential difference, cations will gradually diffuse into the aqueous solution, resulting in the formation of a diffusion electric double layer with uneven cation distribution around the clay. There are exchangeable cations between the crystal layers of clay, which will lead to the swelling of crystal layer spacing after hydration [28]. The main reason is that the repulsion of the clay electric double layer leads to the swelling of the clay lattice. If the amount of lattice swelling is large and the repulsion between clay crystal layers exceeds the attraction between clay crystal layers, the crystal structure will be destroyed. The destruction of crystal structure will reduce the cohesion strength of rock mass and eventually lead to the instability of shaft wall [29]. It can be seen from Figure 12 that surface hydration and osmotic hydration are the scope of the action of clay inhibitors, and their action points are between crystal layers, i.e., small-molecular quaternary ammonium salt inhibitors need to enter into the crystal layers of clay to play a role. Through the adsorption between the primary amine group dissociated by the amine inhibitor and the crystal layer of the active clay, the upper and lower crystal layers of the clay are pulled so as to prevent the crystal layer swelling caused by hydration. Small-molecule quaternary ammonium salt has a polyhydroxy structure, which can provide multiple adsorption sites on the clay's surface and enhance the adsorption of the inhibitor on the clay. The inhibitor was adsorbed on the clay's surface to neutralize the negative charge of the clay or attached to the crystal layer of the clay, reducing the charge between the crystal layer and the surface. At the same time, it can bind the clay through electrostatic adsorption, hydrogen bonding, anchoring and hydrophobic action, and effectively inhibit the hydration, swelling and dispersion of the clay. Usually, the clay surface has a negative charge. When lattice substitution occurs, due to the potential difference and following the principle of minimum energy, an equal amount of opposite charge must be adsorbed on the clay surface, compressing the clay and diffusing the electric double layer. It can inhibit the hydration of clay, thus showing a good inhibition performance in clay.

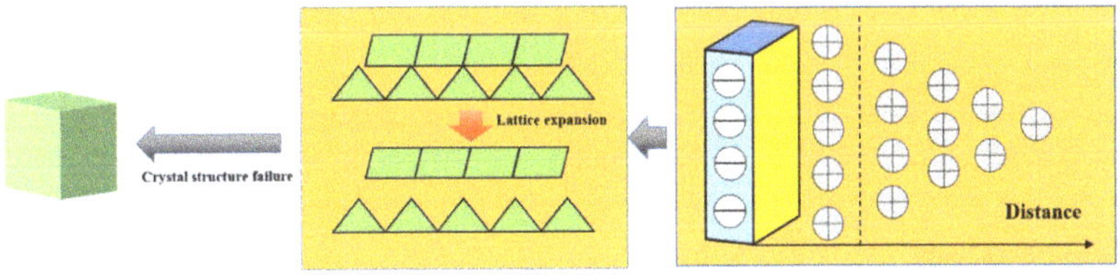

Figure 11. Wellbore instability mechanism 1.

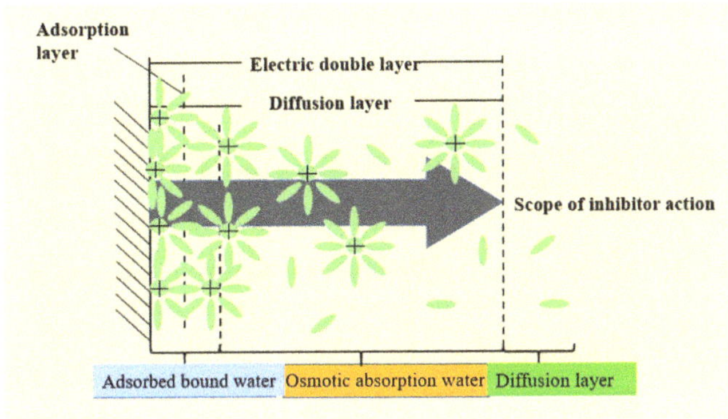

Figure 12. Wellbore instability mechanism 2.

4. Conclusions

In conclusion, a small-molecule quaternary ammonium salt inhibitor was synthesized from triethanolamine and epichlorohydrin. The inhibition performance of 0.5% inhibitor TEE-2 was studied by FTIR spectroscopy and performance evaluation experiments. The results show that 0.5% TEE-2 had an obvious inhibitory effect on hydration. The linear swelling rate of 0.5% TEE-2 was only 29.34%, which was 32.97% and 15.82% lower than that of tap water and 4% KCl, respectively. The performance of inhibitors was also studied by particle size analysis, SEM and TGA. The synthesized inhibitor had the structure of small polyhydroxy molecules, and it can provide multiple adsorption sites on a clay surface and enhance the adsorption of the inhibitor in clay. The inhibitor was adsorbed on the clay's surface and the clay was negatively charged or attached to the clay crystal layer to reduce the charge between the crystal layer and the surface. At the same time, the binding of clay was realized through electrostatic adsorption, hydrogen bonding, anchoring, and hydrophobic action, which can effectively inhibit the hydration, swelling and dispersion of clay, so as to show a good inhibition performance in clay.

Author Contributions: Conceptualization, M.S., G.C. and X.G.; methodology, M.S and X.G.; software, H.H. and Y.Z.; validation, J.R. and R.J.; formal analysis, Q.W., H.H., Y.Z. and R.J.; investigation, Q.W., H.H. and Y.Z.; resources, G.C. and X.G.; data curation, Q.W. and H.H.; writing—original draft preparation, Q.W., G.C. and X.G.; writing—review and editing, M.S., G.C. and X.G.; Supervision, M.S. and X.G.; project administration, G.C. and X.G.; funding acquisition, G.C. All authors have read and agreed to the published version of the manuscript.

Funding: This work is funded by Shaanxi Key Research and Development Plan (2023-YBGY-052), Key Scientific Research Project of Shaanxi Provincial Department of Education (21JY035).

Data Availability Statement: The data presented in this study are openly available in article.

Acknowledgments: The authors are grateful for the support of the Youth Innovation Team of Shaanxi University and the work of Modern Analysis and Testing Center of Xi'an Shiyou University.

Conflicts of Interest: The all authors declare no conflicts of interest.

References

1. Quainoo, A.K.; Negash, B.M.; Bavoh, C.B.; Ganat, T.O.; Tackie-Otoo, B.N. Aperspective on the potential application of bio-inhibitors for shale stabilization during drilling and hydraulic fracturing processes. *Nat. Gas Sci. Eng.* **2020**, *79*, 103380. [CrossRef]
2. Muhammed, N.S.; Olayiwola, T.; Elkatatny, S. A review on clay chemistry, characterization and shale inhibitors for water-based drilling fluids. *J. Pet. Sci. Eng.* **2021**, *206*, 109043. [CrossRef]
3. Gholami, R.; Elochukwu, H.; Fakhari, N.; Sarmadivaleh, M. A review on borehole instability in active shale formations: Interactions, mechanisms and inhibitors. *Earth Sci. Rev.* **2018**, *177*, 2–13. [CrossRef]
4. Rana, A.; Arfaj, M.K.; Saleh, T.A. Advanced developments in shale inhibitors for oil production with low environmental footprints—A review. *Fuel* **2019**, *247*, 237–249. [CrossRef]
5. Li, M.C.; Wu, Q.; Han, J.; Mei, C.; Lei, T.; Lee, S.-Y.; Gwon, J. Overcoming salt contamination of bentonite water-based drilling fluids with blended dual functionalized cellulose nanocrystals. *ACS Sustain. Chem. Eng.* **2020**, *8*, 11569–11578. [CrossRef]
6. Li, X.; Jiang, G.; Yang, L.; Wang, K.; Shi, H.; Li, G.; Wu, X. Application of gelatin quaternary ammonium salt as an environmentally friendly shale inhibitor for water-based drilling fluids. *Energy Fuels* **2019**, *33*, 9342–9350. [CrossRef]
7. Wang, Q.N.; Jiang, G.C.; Wang, X.X.; He, Y.; Yang, L.; Wang, Y. Study on polyarginyl lysine grafted modified nano biochar as an environmentally friendly shale inhibitor. *J. Mol. Liq.* **2024**, *410*, 125666. [CrossRef]
8. Ma, J.Y.; Xia, B.R.; An, Y.X. Advanced developments in low-toxic and environmentally friendly shale inhibitor: A review. *J. Pet. Sci. Eng.* **2022**, *208*, 109578.
9. Zhang, C.; Lv, K.H.; Gong, J.Q.; Wang, Z.; Huang, X.; Sun, J.; Yao, X.; Liu, K.; Rong, K.; Li, M. Synthesis of a hydrophobic quaternary ammonium salt as a shale inhibitor for water-based drilling fluids and determination of the inhibition mechanism. *J. Mol. Liq.* **2022**, *362*, 119474. [CrossRef]
10. Wu, J.K.; Xu, H.Y.; Xie, Y.N.; Zhou, X.Y. A research and application of strong inhibition drilling fluid. *Appl. Mech. Mater.* **2014**, *675*, 1481–1484.
11. Han, Y.Z.; Song, Z.Q.; Huang, W.A.; Cao, J. Shale inhibitive properties of polyether diamine in water-based drilling fluid. *Pet. Sci. Technol.* **2011**, *78*, 510–515.
12. Qu, Y.Z.; Lai, X.Q.; Zou, L.F.; Su, Y.N. Polyoxyalkyleneamine as shale inhibitor in water-based drilling fluids. *Appl. Clay Sci.* **2009**, *44*, 264–265. [CrossRef]
13. Wang, J.H.; Yan, J.N.; Ding, T.W. Progresses in the researches on high performance water base muds. *Drill. Fluid Complet. Fluid* **2007**, *1*, 1–2.
14. Ni, W.J.; Wang, W.L.; Wang, Q.C.; Du, W.C.; Chen, G. Modification and application of waste shasdock peel as a green additive for water-based drilling fluid. *J. Bio-Based Mater. Bioenergy* **2021**, *15*, 380–384. [CrossRef]
15. Huang, S.M.; Jiang, G.C.; He, Y.B.; Dong, T.F.; Yang, L.L.; Li, X.; Huang, W.A. Performance evaluation of trimethylolpropane ester as high-temperature resistant lubricant for high performance water-based drilling fluids. *Geoenergy Sci. Eng.* **2024**, *238*, 212868. [CrossRef]
16. Yang, J.; Sun, J.S.; Wang, R.; Qu, Y.Z.; Liu, F.; Feng, X.H.; Wang, H.H.; Yang, Z.X.; Ren, H.; Liu, L.M. Investigation of TPEG comb polymer as filtration and rheological additives for high-temperature water-based drilling fluid. *J. Mol. Liq.* **2024**, *400*, 124577. [CrossRef]
17. Mobeen, M.; Afeez, G.; Syed, M.S.H.; Sulaiman, A.A.; Mohamed, M.; Shirish, P.; Muhammad, S.K. Experimental investigation of pyrrolidinium-based ionic liquid as shale swelling inhibitor for water-based drilling fluids. *Geoenergy Sci. Eng.* **2023**, *231*, 212374.
18. Du, W.C.; Wang, X.Y.; Bi, T.F.; Liu, M.; Zhang, J.; Chen, G. Synthesis and inhibitive mechanism of a novel clay hydration inhibitor for water-based drilling fluids. *Mater. Sci. -Medzg.* **2021**, *27*, 128–142. [CrossRef]
19. Xu, J.G.; Wang, L.; Hu, H.; Cao, D.S.; Li, S.H. Improving shale hydration inhibition with hydrophobically modified graphene oxide in water-based drilling fluids. *J. Mol. Liq.* **2024**, *413*, 125908. [CrossRef]
20. Xie, C.Y.; Jia, N.; He, L.W. Study on the instability mechanism and grouting reinforcement repair of large-scale underground stopes. *Adv. Civ. Eng.* **2020**, *2020*, 1–10. [CrossRef]
21. Zhang, B.W.; Wang, Q.C.; Wei, Y.; Wei, W.; Du, W.C.; Zhang, J.; Chen, G.; Michal, S. Preparation and swelling inhibition of mixed metal hydroxide to bentonite clay. *Minerals* **2022**, *12*, 459. [CrossRef]
22. Xie, G.; Luo, Y.J.; Wang, C.L.; Deng, M.Y.; Bai, Y. Comparative study on the inhibiting mechanism of inhibitor with primary amine groups and quaternary ammonium groups for sodium bentonite. *Petroleum* **2023**, *9*, 526–533. [CrossRef]
23. Wang, Q.D.; Michal, S.; Gu, X.F.; Miao, Z.P.; Du, W.C.; Zhang, J.; Chen, G. Lubricity and rheological properties of highly dispersed graphite in clay-water-based drilling fluids. *Materials* **2022**, *15*, 1083. [CrossRef]
24. Shen, X.L.; Jiang, G.C.; Li, X.L.; He, Y.B.; Yang, L.L.; Cui, K.X.; Li, W.Q. Application of carboxylated cellulose nanocrystals as eco-friendly shale inhibitors in water-based drilling fluids. *Colloids Surf. A Physicochem. Eng. Asp.* **2021**, *627*, 127182. [CrossRef]

25. Liu, F.; Zhang, C.X.; Li, X.Q.; Zhang, Z.X.; Wang, X.W.; Dai, X.D.; Zhou, M.Y.; Liu, Q.X. Investigation of the inhibition mechanism of polymer/nano-silica composite as shale inhibitor in water-based drilling fluids. *Colloids Surf. A Physicochem. Eng. Asp.* **2022**, *636*, 128099. [CrossRef]
26. Su, J.L.; Liu, M.Q.; Lin, L.; Pu, X.L.; Ge, C.M.; Zhang, T.X.; Liu, G.W. Sulfonated lignin modified with silane coupling agent as biodegradable shale inhibitor in water-based drilling fluid. *J. Pet. Sci. Eng.* **2022**, *208*, 109618. [CrossRef]
27. Yao, X.Y.; Liu, K.C.; Zhou, Z.N.; Zhou, J.; Huang, X.B.; Lu, T.M.; Yu, Y.S.; Li, H. Hydrophobic small-molecule polymers as high-temperature-resistant inhibitors in water-based drilling fluids. *Fluid Dyn. Mater. Process.* **2023**, *19*, 1775–1787. [CrossRef]
28. Vryzas, Z.; Kelessidis, V.C. Nano-based drilling fluids: A review. *Energies* **2017**, *10*, 540. [CrossRef]
29. Xuan, Y.; Jiang, G.C.; Li, Y.Y.; Wang, J.; Geng, H. Inhibiting effect of dopamine adsorption and polymerization on hydrated swelling of montmorillonite. *Colloids Surf. A* **2013**, *422*, 50–60. [CrossRef]

MDPI AG
Grosspeteranlage 5
4052 Basel
Switzerland
Tel.: +41 61 683 77 34

Minerals Editorial Office
E-mail: minerals@mdpi.com
www.mdpi.com/journal/minerals

www.ingramcontent.com/pod-product-compliance
Lightning Source LLC
Chambersburg PA
CBHW061606210926
51850CB00009B/2228